Mind and Nature ◧

MIND AND NATURE

Selected Writings on Philosophy, Mathematics, and Physics

HERMANN WEYL

Edited and with an introduction by Peter Pesic

PRINCETON UNIVERSITY PRESS *Princeton and Oxford*

Published by Princeton University Press, 41 William Street, Princeton, New Jersey 08540

In the United Kingdom: Princeton University Press, 6 Oxford Street, Woodstock, Oxfordshire OX20 1TW

Library of Congress Cataloging-in-Publication Data

Weyl, Hermann, 1885–1955.
Mind and nature : selected writings on philosophy, mathematics, and physics / Hermann Weyl; edited and with an introduction by Peter Pesic.
     p.   cm.
  Includes bibliographical references and index.
  ISBN 978-0-691-13545-8 (cloth : alk. paper)   1. Mathematics—Philosophy.
2. Physics—Philosophy.   I. Pesic, Peter.   II. Title.
  QA8.4.W49 2009
  510.1—dc22   2008049964

British Library Cataloging-in-Publication Data is available

This book has been composed in Adobe Garamond
Printed on acid-free paper. ∞
press.princeton.edu
Printed in the United States of America

10 9 8 7 6 5 4 3 2 1

Contents ▣

Mind and Nature ▣

Introduction ▣

"It's a crying shame that Weyl is leaving Zurich. He is a great master."[1] Thus Albert Einstein described Hermann Weyl (1885–1955), who remains a legendary figure, "one of the greatest mathematicians of the first half of the twentieth century.... No other mathematician could claim to have initiated more of the theories that are now being explored," as Michael Atiyah put it.[2] Weyl deserves far wider renown not only for his importance in mathematics and physics but also because of his deep philosophical concerns and thoughtful writing. To that end, this anthology gathers together some of Weyl's most important general writings, especially those that have become unavailable, have not previously been translated into English, or were unpublished. Together, they form a portrait of a complex and fascinating man, poetic and insightful, whose "vision has stood the test of time."[3]

This vision has deeply affected contemporary physics, though Weyl always considered himself a mathematician, not a physicist.[4] The present volume emphasizes his treatment of philosophy and physics, but another complete anthology could be made of Weyl's general writings oriented more directly toward mathematics. Here, I have chosen those writings that most accessibly show how Weyl synthesized philosophy, physics, and mathematics.

Weyl's philosophical reflections began in early youth. He recollects vividly the worn copy of a book about Immanuel Kant's *Critique of Pure Reason* he found in the family attic and read avidly at age fifteen. "Kant's teaching on the 'ideality of space and time' immediately took powerful hold of me; with a jolt I was awakened from my 'dogmatic slumber,' and the mind of the boy

1

found the world being questioned in radical fashion." At the same time, he was drinking deep of great mathematical works. As "a country lad of eighteen," he fell under the spell of David Hilbert, whom he memorably described as a Pied Piper "seducing so many rats to follow him into the deep river of mathematics"; the following summer found Weyl poring over Hilbert's *Report on the Theory of Algebraic Numbers* during "the happiest months of my life."[5] As these stories reveal, Weyl was a very serious man; Princeton students called him "holy Hermann" among themselves, mocking a kind of earnestness probably more common in Hilbert's Göttingen. There, under brilliant teachers who also included Felix Klein and Hermann Minkowski, Weyl completed his mathematical apprenticeship. Forty years later, at the Princeton Bicentennial in 1946, Weyl gave a personal overview of this period and of the first discoveries that led him to find a place of distinction at Hilbert's side. This address, never before published, may be a good place to begin if you want to encounter the man and hear directly what struck him most. Do not worry if you find the mathematical references unfamiliar; Weyl's tone and angle of vision express his feelings about the mathematics (and mathematicians) he cared for in unique and evocative ways; he describes "Koebe the rustic and Brouwer the mystic" and the "peculiar gesture of his hands" Koebe used to define Riemann surfaces, for which Weyl sought "a more dignified definition."

In this address, Weyl also vividly recollects how Einstein's theory of general relativity affected him after the physical and spiritual desolation he experienced during the Great War. "In 1916 I had been discharged from the German army and returned to my job in Switzerland. My mathematical mind was as blank as any veteran's and I did not know what to do. I began to study algebraic surfaces; but before I had gotten far, Einstein's memoir came into my hands and set me afire."[6] Both Weyl and Einstein had lived in Zurich and taught at its Eidgenössische Technische Hochschule (ETH) during the very period Einstein was struggling to find his generalized theory, for which he needed mathematical help.[7] This was a golden period for both men, who valued the freer spirit they found in Switzerland, compared to Germany. Einstein adopted Swiss citizenship, completed his formal education in his new country, and then worked at its patent office. Among the first to realize the full import of Einstein's work, especially its new, more general, phase, Weyl gave lectures on it at the ETH in 1917, published in his eloquent book *Space-Time-Matter* (1918). Not only the first (and perhaps the greatest) extended account of Einstein's general relativity, Weyl's book was immensely influential because of its profound sense of perspective, great expository clarity, and indications of directions to carry Einstein's work further. Einstein himself praised the book as a "symphonic masterpiece."[8]

As the first edition of *Space-Time-Matter* went to press, Weyl reconsidered Einstein's ideas from his own mathematical perspective and came upon a new and intriguing possibility, which Einstein immediately called "*a first-class stroke of genius.*"[9] Weyl describes this new idea in his essay "Electricity and Gravitation" (1921), much later recollecting some interesting personal details in his Princeton address. There, Weyl recalls explaining to a student, Willy Scherrer, "that vectors when carried around by parallel displacement may return to their starting point in changed direction. And he asked 'Also with changed length?' Of course I gave him the orthodox answer [no] at that moment, but in my bosom gnawed the doubt." To be sure, Weyl wrote this remembrance thirty years later, which thus may or may not be a perfectly faithful record of the events; nevertheless, it represents Weyl's own self-understanding of the course of his thinking, even if long after the fact. Though Weyl does not mention it, this conversation was surrounded by a complex web of relationships: Weyl's wife Helene was deeply involved with Willy's brother Paul, while Weyl himself was the lover of Erwin Schrödinger's wife, Anny. These personal details are significant here because Weyl himself was sensitive to the erotic aspects of scientific creativity in others, as we will see in his commentary on Schrödinger, suggesting that Weyl's own life and works were similarly intertwined.[10]

In *Space-Time-Matter*, Weyl used the implications of parallel transport of vectors to illuminate the inner structure of the theory Einstein had originally phrased in purely *metric* terms, meaning the measurement of distances between points, on the model of the Pythagorean theorem.[11] Weyl questioned the implicit assumption that behind this metrical structure is a fixed, given distance scale, or "gauge." What if the direction as well as the length of meter sticks (and also the standard second given by clocks) were to vary at different places in space-time, just as railway gauge varies from country to country? Perhaps Weyl's concept began with this kind of homely observation about the "gauge relativity" in the technology of rail travel, well-known to travelers in those days, who often had to change trains at frontiers between nations having different, incompatible railway gauges.[12] By considering this new kind of relativity, Weyl stepped even beyond the general coordinate transformations Einstein allowed in his general theory so as to incorporate what Weyl called *relativity of magnitude*.

In what Weyl called an "affinely connected space," a vector could be displaced parallel to itself, at least to an infinitely nearby point. As he realized after talking with Willy Scherrer, in such a space a vector transported around a closed curve might return to its starting point with changed direction *and* length (which he called "non-integrability," as measured mathematically by the "affine connection"). Peculiar as this changed length might seem, Weyl was struck by the mathematical generality of this possibility, which he explored in what he called his "purely infinitesimal geometry," which emphasized infinitesimal displacement as the foundation in terms of which any finite displacement needed to be understood.[13] As he emphasized the centrality of the infinite in mathematics, Weyl also placed the infinitesimal before the finite.

Weyl also realized that his generalized theory gave him what seemed a natural way to incorporate electromagnetism into the structure of space-time, a goal that had eluded Einstein, whose theory treated electromagnetism along with matter as mass-energy *sources* that caused space-time curvature but remained separate from space-time itself. Here Weyl used the literal "gauging" of distances as the basis of a mathematical analogy; his reinterpretation of these equations led naturally to a *gauge field* he could then apply to electromagnetism, from which Maxwell's equations now emerged as intrinsic to the structure of space-time. Though Einstein at first hailed this "stroke of genius," soon he found what he considered a devastating objection: because of the non-integrability of Weyl's gauge field, atoms would not produce the constant, universal spectral lines we actually observe: atoms of hydrogen on Earth give the same spectrum as hydrogen observed telescopically in distant stars. Weyl's 1918 paper announcing his new theory appeared with an unusual postscript by Einstein, detailing his objection, along with Weyl's reply that the actual behavior of atoms in turbulent fields, not to speak of measuring rods and clocks, was not yet fully understood. Weyl noted that his theory used light signals as a fundamental standard, rather than relying on supposedly rigid measuring rods and idealized clocks, whose atomic structure was in some complex state of accommodation to ambient fields.[14]

In fact, the atomic scale was the arena in which quantum theory was then emerging. Here began the curious migration of Weyl's idea from literally regauging length and time to describing some other realm beyond space-time. Theodor Kaluza (1922) and Oskar Klein (1926) proposed a generalization of general relativity using a fifth dimension to accommodate electromagnetism. In their theory, Weyl's gauge factor turns into a *phase* factor, just as the relative phase of traveling waves depends on the varying dispersive properties of the medium they traverse. If so, Weyl's gauge would no longer be immediately observable (as Einstein's objection asserted) because the gauge affects only the *phase*, not the observable *frequency*, of atomic spectra.[15]

At first, Weyl speculated that his 1918 theory gave support to the radical possibility that "matter" is only a form of curved, empty space (a view John Wheeler championed forty years later). Here Weyl doubtless remembered the radical opinions of Michael Faraday and James Clerk Maxwell, who went so far as to consider so-called matter to be a nexus of immaterial lines of force.[16]

Weyl then weighed these mathematical speculations against the complexities of physical experience. Though he still believed in his fundamental insight that gauge invariance was crucial, by 1922 Weyl realized that it needed to be reconsidered in light of the emergent quantum theory. Already in 1922, Schrödinger pointed out that Weyl's idea could lead to a new way to understand quantization. In 1927, Fritz London argued that the gravitational scale factor implied by Weyl's 1918 theory, which Einstein had argued was unphysical, actually makes sense as the complex phase factor essential to quantum theory.[17]

As Schrödinger struggled to formulate his wave equation, at many points he relied on Weyl for mathematical help. In their liberated circles, Weyl remained a valued friend and colleague even while being Anny Schrödinger's lover. From that intimate vantage point, Weyl observed that Erwin "did his great work during a late erotic outburst in his life," an intense love affair simultaneous with Schrödinger's struggle to find a quantum wave equation. But then, as Weyl inscribed his 1933 Christmas gift to Anny and Erwin (a set of erotic illustrations to Shakespeare's *Venus and Adonis*), "The sea has bounds but deep desire has none."[18]

Weyl's insight into the nature of quantum theory comes forward in a pair of letters he and Einstein wrote in 1922, here reprinted and translated for the first time, responding to a journalist's question about the significance of the new physics. Einstein dismisses the question: for him in 1922, relativity theory changes nothing fundamental in our view of the world, and that is that. Weyl takes the question more seriously, finding a radically new insight not so much in relativity theory as in the emergent quantum theory, which Weyl already understood as asserting that "the entire physics of matter is statistical in nature," showing how clearly he understood this decisive point several years before the formulation of the new quantum theory in 1925–1926 by Max Born, Werner

Heisenberg, Pascual Jordal, P.A.M. Dirac, and Schrödinger. In his final lines, Weyl also alludes to his view of matter as agent (*agens*), in which he ascribed to matter an innate activity that may have helped him understand and accept the spontaneity and indeterminacy emerging in quantum theory. This view led Weyl to reconsider the significance of the concept of a field. As he wrote to Wolfgang Pauli in 1919, "field physics, I feel, really plays only the role of 'world geometry'; in matter there resides still something different, [and] real, that cannot be grasped causally, but that perhaps should be thought of in the image of 'independent decisions,' and that we account for in physics by statistics."[19]

In the years around 1920, Weyl continued to work out the consequences of this new approach. His conviction about the centrality of consciousness as intuition and activity deeply influenced his view of matter. As the ego drives the whole world known to consciousness, he argued that "matter is analogous to the ego, the effects of which, despite the ego itself being non-spatial, originate via its body at a given point of the world continuum. Whatever the nature of this *agens*, which excites the field, might be—perhaps life and will—in physics we only look at the field effects caused by it." This took him in a direction very different from the vision of matter reduced to pure geometry he had entertained in 1918. Writing to Felix Klein in 1920, Weyl noted that "field physics no longer seems to me to be the key to reality; but the field, the ether, is to me only a totally powerless *transmitter* by itself of the action, but matter rests beyond the field and is the reality that *causes* its states." Weyl described his new view in 1923 using an even more striking image: "Reality does not move into space as into a right-angled tenement house along which all its changing play of forces glide past without leaving any trace; but rather as the snail, matter itself builds and shapes this house of its own." For Weyl now, fields were "totally powerless *transmitters*" that are not really existent or effectual in their own right, but only a way of talking about states of *matter* that are the locus of fundamental reality. Though Weyl still retained fields to communicate interactions, his emphasis that the reality of "matter rests beyond the field" may have influenced Richard Feynman and Wheeler two decades later in their own attempt to remove "fields" as independent beings. Weyl also raised the question of whether matter has some significant topological structure on the subatomic scale, as if topology were a kind of activation that brings static geometry to life, analogous to the activation the ego infuses into its world. Such topological aspects of matter only emerged as an important frontier of contemporary investigation fifty years later.[20]

Looking back from 1955 at his original 1918 paper, Weyl noted that he "had no doubt" that the correct context of his vision of gauge theory was "not, as I believed in 1918, in the intertwining of electromagnetism and gravity" but in "the Schrödinger-Dirac potential ψ of the electron-positron field. . . . The strongest argument for my theory seems to be this, that gauge-invariance corresponds to the conservation of electric charge in the same way that coordinate-invariance corresponds to the conservation of energy and momentum," the insight that Emmy Noether's famous theorem put at the foundations of quantum field theory.[21] Nor did Weyl himself stop working on his idea; in 1929 he published an important paper reformulating his idea in the language of what today are called gauge fields; these considerations also led him to consider fundamental physical symmetries long before the discovery of the violation of parity in the 1950s. The "Weyl two-component neutrino field" remains a standard description of neutrinos, all of which are "left-handed" (spin always opposed to direction of motion), as all antineutrinos are "right-handed." In 1954 (a year before Weyl's death, but apparently not known to him), C. N. Yang, R. Mills, and others took the next steps in developing gauge fields, which ultimately became the crucial element in the modern "standard model" of particle physics that triumphed in the 1970s, unifying strong, weak, and electromagnetic interactions in ways that realized Weyl's distant hopes quite beyond his initial expectations.[22]

In the years that Weyl continued to try to find a way to make his idea work, he and Einstein underwent a curious exchange of positions. Originally, Einstein thought Weyl was not paying enough attention to physical measuring rods and clocks because Weyl used immaterial light beams to measure space-time. As Weyl recalled in a letter of 1952,

> I thought to be able to answer his concrete objections, but in the end he said: "Well, Weyl, let us stop this. For what I actually have against your theory is: 'It is impossible to do physics like this (i.e., in such a speculative fashion, without a guiding intuitive physical principle)!'" Today we have probably changed our viewpoints in this respect: E. believes that in this domain the chasm between ideas and experience is so large, that only mathematical speculation (whose consequences, of course, have to be developed and confronted with facts) gives promise of success, while my confidence in pure speculation has diminished and a closer connection with quantum physical experience

seems necessary, especially as in my view it is not sufficient to blend gravitation and electricity to one unity, but that the wave fields of the electron (and whatever there may still be of nonreducible elementary particles) must be included.[23]

Ironically, Weyl the mathematician finally sided with the complex realities of physics, whereas Einstein the physicist sought refuge in unified field theories that were essentially mathematical. Here is much food for thought about the philosophic reflections each must have undertaken in his respective soul search- ing and that remain important now, faced with the possibilities and problems of string theory, loop quantum gravity, and other theoretical directions for which sufficient experimental evidence may long remain unavailable.

Both here and throughout his life, Weyl used philosophical reflection to guide his theoretical work, preferring "to approach a question through a deep analysis of the concepts it involves rather than by blind computations," as Jean Dieudonné put it. Though others of his friends, such as Einstein and Schrödinger, shared his broad humanistic education and philosophical bent, Weyl tended to go even further in this direction. As a young student in Göttin- gen, Weyl had studied with Edmund Husserl (who had been a mathematician before turning to philosophy), with whom Helene Weyl had also studied.[24]

Weyl's continuing interest in phenomenological philosophy marks many of his works, such as his 1927 essay on "Time Relations in the Cosmos, Proper Time, Lived Time, and Metaphysical Time," here reprinted and translated for the first time. The essay's title indicates its scope, beginning with his interpre- tation of the four-dimensional space-time Hermann Minkowski introduced in 1908, which Weyl then connects with human time consciousness (also a deep interest of Husserl's). Weyl treats a world point not merely as a mathematical abstraction but as situating a "point-eye," a living symbol of consciousness peering along its world line. Counterintuitively, that point-eye associates the objective with the relative, the subjective with the absolute.

Weyl uses this striking image to carry forward a mathematical insight that had emerged earlier in his considerations about the nature of the continuum. During the early 1920s, Weyl was deeply drawn to L.E.J. Brouwer's advocacy of intuition as the critical touchstone for modern mathematics. Thus, Brouwer rejected Cantor's transfinite numbers as not intuitable, despite Hilbert's claim that "no one will drive us from the paradise which Cantor created for us." Hilbert argued that mathematics should be considered purely formal, a great game in which terms like "points" or "lines" could be replaced with arbitrary words like "beer mugs" or "tables" or with pure symbols, so long as the axiomatic relationships between the respective terms do not change. Was this, then, the "deep river of mathematics" into which Weyl thought this Pied Piper had lured him and so many other clever young rats?[25]

By the mid-1920s, Weyl was no longer an advocate of Brouwer's views (though still reaffirming his own 1918 work on the continuum). In his magisterial *Philosophy of Mathematics and Natural Science* (written in 1927 but extensively revised in 1949), Weyl noted that "mathematics with Brouwer gains its highest intuitive clarity. . . . It cannot be denied, however, that in advancing to higher and more general theories the inapplicability of the simple laws of classical logic eventually results in an almost unbearable awkwardness. And the mathematician watches with pain the larger part of his towering edifice which he believed to be built of concrete blocks dissolve into a mist before his eyes."[26]

Even so, Weyl remained convinced that we should not consider a continuum (such as the real numbers between 0 and 1) as an actually completed and infinite set but only as capable of endless subdivision. This understanding of the "potential infinite" recalls Aristotle's critique of the "actual infinite." In his 1927 essay on "Time Relations," Weyl applied this view to time as a continuum. Because an infinitely small point could be generated from a finite interval only through actually completing an infinite process of shrinkage, Weyl applies the same argument to the presumption that the present instant is a "point in time." He concludes that "there is no pointlike Now and also no exact earlier and later."[27] Weyl's arguments about the continuum have the further implication that the past is *never completely determined*, any more than a finite, continuous interval is ever exhaustively filled; both are potentially infinite because always further divisible. If so, the past is not fixed or unchangeable and continues to change, a luxuriant, ever-proliferating tangle of "world tubes," as Weyl called them, "open into the future and again and again a fragment of it is lived through." This intriguing idea is psychologically plausible: A person's past seems to keep changing and ramifying as life unfolds; the past today seems different than it did yesterday. As a character in Faulkner put it, "the past is never dead. It's not even past."[28]

In Weyl's view, a field is intrinsically continuous, endlessly subdividable, and hence an abyss in which we never come to an ultimate *point* where a decision can be made: To be or not to be? Conversely, pointlike, discrete matter is a locus of decisive spontaneity because it is not predictable through continuous field laws, only observable statistically. As Weyl wrote to Pauli in 1919, "I am firmly convinced that statistics is in principle something independent from causality, the 'law'; because it is in general absurd to imagine a continuum as something like a finished being." Because of this independence, Weyl continued in 1920,

> the future will act on and upon the present and it will determine the present more and more precisely; the past is not finished. Thus, the fixed pressure of natural causality disappears and there remains, irrespective of the validity of the natural laws, *a space for autonomous and causally absolutely independent decisions*; I consider the elementary quanta of matter to be the place of these decisions.[29]

"Lived time," in Weyl's interpretation, keeps evoking the past into further life, even as it calls the future into being. Weyl's deep thoughts may still repay the further exploration they have not received so far.

Weyl also contributed notably to the application of general relativity to cosmology. He found new solutions to Einstein's equation and already in 1923 calculated a value for the radius of the universe of roughly one billion light years, six years before Edwin Hubble's systematic measurements provided what became regarded as conclusive evidence that our galaxy is only one among many. Weyl also reached a seminal insight, derived from both his mathematical and his philosophical considerations, that the topology of the universe is "the first and most important question in all speculations on the world as a whole." This prescient insight was taken up only in the 1970s and remains today at the forefront of cosmology, still unsolved and as important as Weyl thought. He also noted that relativistic cosmology indeed "left the door open for possibilities of every kind." The mysteries of dark energy and dark matter remind us of how much still lies beyond that door. Then too, we still face the questions Weyl raised regarding the strange recurrence throughout cosmology of the "large numbers" like 10^{20} and 10^{40} (seemingly as ratios between cosmic and atomic scales), later rediscovered by Dirac.[30]

Other of Weyl's ideas long ago entered and transformed the mainstream of physics, characteristically bridging the mathematical and physical through the philosophical. He considered his greatest mathematical work the classification of the semisimple groups of continuous symmetries (Lie groups), which he later surveyed in *The Classical Groups: Their Invariants and Representations* (1938).

In the introduction to this first book he wrote in English, Weyl noted that "the gods have imposed upon my writing the yoke of a foreign tongue that was not sung at my cradle." But even in his adopted tongue he does not hesitate to critique the "too thorough technicalization of mathematical research" in America that has led to "a mode of writing which must give the reader the impression of being shut up in a brightly illuminated cell where every detail sticks out with the same dazzling clarity, but without relief. I prefer the open landscape under a clear sky with its depth of perspective, where the wealth of sharply defined nearby details gradually fades away toward the horizon." Such writing exemplifies Weyl's uniquely eloquent style.

Soon after quantum theory had first been formulated, Weyl used his deep mathematical perspective to shape *The Theory of Groups and Quantum Mechanics* (1928). It is hard to overstate the importance of his marriage of the mathematical theory of symmetry to quantum theory, which has proved ever more fruitful, with no end in sight. At first, as eminent and hardheaded a physicist as John Slater resisted the "group-pest" as if it were a plague of abstractness. But Weyl, along with Eugene Wigner, prevailed because the use of group theory gave access to the symmetries essential for formulating all kinds of physical theories, from crystal lattices to multiplets of fundamental particles. It was this depth and generality that moved Julian Schwinger to "read and re-read that book, each time progressing a little farther, but I cannot say that I ever—not even to this day—fully mastered it." Thus, Schwinger considered Weyl "one of my gods," not merely an outstanding teacher, because "the ways of gods are mysterious, inscrutable, and beyond the comprehension of ordinary mortals."[31] This from someone regarded as rather godlike by many physicists because of his own inscrutable powers. Weyl's insights about the fundamental mathematical symmetries led Schwinger and others decades later to formulate the *TCP* theorem, which expresses the fundamental identicality between particles and antiparticles under the combined symmetries of time reversal (T), charge conjugation (reversal of the sign of the charge, C), and parity (mirror) reversal (P).

In one of his most powerful interventions in physics, Weyl used such symmetry principles to argue that Dirac's newly proposed (and as-yet unobserved) "holes" (antielectrons) could not be (as Dirac had suggested) protons, which are almost two thousand times heavier than electrons. Weyl showed mathematically that antielectrons had to have the same mass as electrons, though having opposite charge; this was later confirmed by cosmic ray observations.

Weyl's purely mathematical argument struck Dirac, who drew from this experience his often-cited principle that "it is more important to have beauty in one's equations than to have them fit experiment," a principle that continues to be an important touchstone for many physicists. Even though Weyl's mathematics moved Dirac to this radical declaration, Weyl's own turn away from mathematical speculation about physics raises the question whether in the end to prefer beautiful mathematics to the troubling complexities of experience.[32]

Whether a "god" or no, Weyl seemed to feel that the philosophical enterprise cannot remain on the godlike plane but really requires the occasions of human conversation. The two largest works in this anthology contain the rich harvest of Weyl's long-standing interest in expressing his ideas to a broader audience; both began as lecture series, thus doubly public, both spoken and written. To use the apt phrase of his son Michael, *The Open World* (1932) contains "Hermann's dialogues with God" because here the mathematician confronts his ultimate concerns.[33] These do not fall into the traditional religious traditions but are much closer in spirit to Spinoza's rational analysis of what he called "God or nature," so important for Einstein as well. As Spinoza considered the concept of infinity fundamental to the nature of God, Weyl defines "God as the completed infinite." In Weyl's conception, God is not merely a mathematician but is mathematics itself because "*mathematics is the science of the infinite*," engaged in the paradoxical enterprise of seeking "the symbolic comprehension of the infinite with human, that is finite, means." In the end, Weyl concludes that this God "cannot and will not be comprehended" by the human mind, even though "mind is freedom within the limitations of existence; it is open toward the infinite." Nevertheless, "neither can God penetrate into man by revelation, nor man penetrate to him by mystical perception. The completed infinite we can only represent in symbols." In Weyl's praise of openness, this freedom of the human mind begins to seem even higher than the completed infinity essential to the meaning of God. Does not his argument imply that God, as actual infinite, can never be *actually* complete, just as an infinite time will never have passed, however long one waits? And if God's actuality will never come to pass, in what sense could or does or will God exist at all? Perhaps God, like the continuum or the field, is an infinite abyss that needs completion by the decisive seed of matter, of human choice.

Weyl inscribes this paradox and its possibilities in his praise of the *symbol*, which includes the mathematical no less than the literary, artistic, poetic, thus bridging the presumed chasm between the "two cultures." At every turn in his writing, we encounter a man of rich and broad culture, at home in many domains of human thought and feeling, sensitive to its symbols and capable of expressing himself beautifully. He moves so naturally from quoting the ancients and moderns to talking about space-time diagrams, thus showing us something of his innate turn of mind, his peculiar genius. His quotations and reflections are not mere illustrations but show the very process by which his thought lived and moved. His philosophical turn of mind helped him reach his own finest scientific and mathematical ideas. His self-deprecating disclaimer that he thus "wasted his time" might be read as irony directed to those who misunderstood him, the hardheaded who had no feeling for these exalted ideas and thought his philosophizing idle or merely decorative. Weyl gained perspective, insight, and altitude by thinking back along the ever-unfolding past and studying its great thinkers, whom he used to help him soar, like a bird feeling the air under its wings.

In contrast, Weyl's lectures on *Mind and Nature*, published only two years later (1934), have a less exalted tone. The difference shows his sensitivity to the changing times. Though invited to return to Göttingen in 1918, he preferred to remain in Zurich; finally in 1930, he accepted the call to succeed Hilbert, but almost immediately regretted it. The Germany he returned to had become dangerous for him, his Jewish wife, and his children. Unlike some who were unable to confront those ugly realities, Weyl was capable of political clear-sightedness; by 1933 he was seeking to escape Germany. His depression and uncertainty in the face of these huge decisions shows another side of his humanity; as Richard Courant put it, "Weyl is actually in spite of his enormously broad talents an inwardly insecure person, for whom nothing is more difficult than to make a decision which will have consequences for his whole life, and who mentally is not capable of dealing with the weight of such decisions, but needs a strong support somewhere."[34] That anxiety and inner insecurity gives Weyl's reflections their existential force. As he himself struggled along his own world line through endlessly ramifying doubts, he came to value the spontaneity and decisiveness he saw in the material world.

Weyl's American lectures marked the start of a new life, beginning with a visiting professorship at Princeton (1928–1929), where he revised his book on group theory and quantum mechanics in the course of introducing his insights to this new audience. Where in 1930 Weyl's *Open World* began with God, in 1933 his lectures on *Mind and Nature* start with human subjectivity and sense perception. Here, symbols help us confront a world that "does not exist in itself, but is merely encountered by us as an object in the correlative variance of subject and object." For Weyl, mathematical and poetic symbols may disclose a path through the labyrinth of "mirror land," a world that may seem ever more distorted, unreal on many fronts. Though Weyl discerns "an abyss which no realistic conception of the world can span" between the physical processes of the brain and the perceiving subject, he finds deep meaning in "the enigmatic twofold nature of the ego, namely that I am both: on the one hand a real individual which performs real psychical acts, the dark, striving and erring human being that is cast out into the world and its individual fate; on the other hand light which beholds itself, intuitive vision, in whose consciousness that is pregnant with images and that endows with meaning, the world opens up. Only in this 'meeting' of consciousness and being both exist, the world and I."[35]

Weyl treats relativity and quantum theory as the latest and most suggestive symbolic constructions we make to meet the world. The dynamic character of symbolism endures, even if the particular symbols change; "their truth refers to a connected system that can be confronted with experience only as a whole."[36] Like Einstein, Weyl emphasized that physical concepts as symbols "are constructions within a free realm of possibilities," freely created by the human mind. "Indeed, space and time are nothing in themselves, but only a certain order of the reality existing and happening in them." As he noted in 1947, "it has now become clear that physics needs no such ultimate objective entities as space, time, matter, or 'events,' or the like, for its constructions symbols without meaning handled according to certain rules are enough."[37] In *Mind and Nature*, Weyl notes that "in nature itself, as [quantum] physics constructs it theoretically, the dualism of object and subject, of law and freedom, is already most distinctly predesigned." As Niels Bohr put it, this dualism rests on "the old truth that we are both spectators and actors in the great drama of existence."

After Weyl left Germany definitively for Princeton in 1933, he continued to reflect on these matters. In the remaining selections, one notes him retelling some of the same stories, quoting the same passages from great thinkers of the past, repeating an idea he had already said elsewhere. These repetitions posed a difficult problem, for the later essays contain some interesting new points along with the old. Because of this, I decided to include these later essays, for Weyl's repetitions also show him reconsidering. Reiterating a point in a new or larger context may open further dimensions. Then too, we as readers are given another chance to think about Weyl's points and also see where he held to his earlier ideas and where he may have changed. For he *was* capable of changing his mind, more so than Einstein, whose native stubbornness may well have contributed to his unyielding resistance to quantum theory. As noted above, Weyl was far more able to entertain and even embrace quantum views, despite their strangeness, precisely because of his philosophical openness.[38]

Weyl's close reading of the past and his philosophical bent inspired his continued openness. In his hitherto unpublished essay "Man and the Foundations of Science" (written about 1949), Weyl describes "an ocean traveler who distrusts the bottomless sea and therefore clings to the view of the disappearing coast as long as there is in sight no other coast toward which he moves. I shall now try to describe the journey on which the old coast has long since vanished below the horizon. There is no use in staring in that direction any longer."[39] He struggles to find a way to speak about "a new coast [that] seems dimly discernible, to which I can point by dim words only, and maybe it is merely a bank of fog that deceives me." Here symbols might be all we have, for "it becomes evident that now the words 'in reality' must be put between quotation marks; we have a symbolic construction, but nothing which we could seriously pretend to be the true real world." Yet even legerdemain with symbols cannot hide the critical problem of the continuum: "The sin committed by the set-theoretic mathematician is his treatment of the field of possibilities open into infinity as if it were a completed whole all members of which are present and can be overlooked with one glance. For those whose eyes have been opened to the problem of infinity, the majority of his statements carry no meaning. If the true aim of the mathematician is to master the infinite by finite means, he has attained it by fraud only—a gigantic fraud which, one must admit, works as beautifully as paper money." By his reaffirmation of his critique of the actual infinite, we infer that Weyl continued to hold his radical views about "lived time," especially that "we stand at that intersection of bondage and freedom which is the essence of man himself."

Indeed, Weyl notes that he had put forward this relation between being and time years before Martin Heidegger's famous book on that subject appeared. Weyl's account of Heidegger is especially interesting because of the intersection between their concerns, no less than their deep divergences. Yet Weyl seemingly could not bring himself to give a full account of Heidegger or of his own reactions, partly based on philosophical antipathy, partly (one infers) from his profound distaste for Heidegger's involvement with the Nazi regime. Though he does not speak of it, Weyl may well also have known of the way Heidegger abandoned their teacher, Husserl, in those dark days. Most of all, Weyl conveys his annoyance that Heidegger had botched important ideas that were important to Weyl himself and, in the process, that Heidegger had lost sight of the nature of science. Taking up a crucial term they both use, Weyl asserts that "no other ground is left for science to build on than this dark but very solid rock which I once more call the concrete Dasein of man in his world." Weyl grounds this Dasein, man's being-in-the-world, in ordinary language, which is "neither tarnished poetry nor a blurred substitute for mathematical symbolism; on the contrary, neither the one nor the other would and could exist without the nourishing stem of the language of our everyday life, with all it complexity, obscurity, crudeness, and ambiguity." By thus connecting mathematical and poetic symbolism as both growing from the soil of ordinary human language, Weyl implicitly rejects Heidegger's turn away from modern mathematical science.

In his late essay "The Unity of Knowledge" (1954), Weyl reviews this ground and concludes that "the shield of Being is broken beyond repair," but does not take this disunity in a tragic sense because "on the side of Knowing there may be unity. Indeed, mind in the fullness of its experience has unity. Who says 'I' points to it." Here he reaffirms his old conviction that human consciousness is not simply the product of other, more mechanical forces, but is itself the luminous center constituting that reality through its "complex symbolic creations which this lumen built up in the history of mankind." Even though "myth, religion, and alas! also philosophy" fall prey to "man's infinite capacity for self-deception," Weyl implicitly holds out greater hope for the symbolic creations of mathematics and science, though he admits that he is still struggling to find clarity.

The final essay in this anthology, "Insight and Reflection" (1955), is Weyl's rich *Spätlese*, the intense, sweet wine made from grapes long on the vine. This philosophical memoir discloses his inner world of reflection in ways his other, earlier essays did not reveal quite so directly, perhaps aware of the skepticism and irony that may have met them earlier on. We are reminded of his "point-eye," disclosing his thoughts and feelings while creeping up his own world line. Nearing its end, Weyl seems freer to say what he feels, perhaps no longer caring who might mock. He gives his fullest avowal yet of what Husserl meant to him, but does not hold back his own reservations; Husserl finally does not help with Weyl's own deep question about "the relation between the one pure

I of immanent consciousness and the particular lost human being which I find myself to be in a world full of people like me (for example, during the afternoon rush hour on Fifth Avenue in New York)." Weyl is intrigued by Fichte's mystic strain, but in the end Fichte's program (analyzing everything in terms of I and not-I) strikes him as "preposterous." Weyl calls Meister Eckhart "the deepest of the Occidental mystics . . . a man of high responsibility and incomparably higher nobility than Fichte." Eckhart's soaring theological flight beyond God toward godhead stirred Weyl, along with Eckhart's fervent simplicity of tone. Throughout his account, Weyl interweaves his mathematical work, his periods of soberness after the soaring flights of philosophical imagination, though he presents them as different sides of what seems to his point-eye a unified experience. Near the end, he remembers with particular happiness his book *Symmetry* (1952), which so vividly unites the poetic, the artistic, the mathematical, and the philosophical, a book no reader of Weyl should miss.[40] In quoting T. S. Eliot that "the world becomes stranger, the pattern more complicated," we are aware of Weyl's faithful openness to that strangeness, as well as the ever more complex and beautiful symmetries he discerned in it.

Weyl's book on symmetry shows the fundamental continuity of themes throughout his life and work. Thinking back on the theory of relativity, Weyl describes it not (as many of his contemporaries had) as disturbing or revolutionary but really as "another aspect of symmetry" because "it is the inherent symmetry of the four-dimensional continuum of space and time that relativity deals with." Yet as beautifully as he evokes and illustrates the world of symmetry, Weyl still emphasizes the fundamental difference between perfect symmetry and life, with its spontaneity and unpredictability. "If nature were all lawfulness then every phenomenon would share the full symmetry of the universal laws of nature as formulated by the theory of relativity. The mere fact that this is not so proves that *contingency* is an essential feature of the world." Characteristically, Weyl recalls the scene in Thomas Mann's *Magic Mountain*

in which his hero, Hans Castorp, nearly perishes when he falls asleep with exhaustion and leaning against a barn dreams his deep dream of death and love. An hour before when Hans sets out on his unwarranted expedition on skis he enjoys the play of the flakes "and among these myriads of enchanting little stars," so he philosophizes, "in their hidden splendor, too small for man's naked eye to see, there was not one like unto another; an endless inventiveness governed the development and unthinkable differentiation of one and the same basic scheme, the equilateral, equiangular hexagon. Yet each in itself—this was the uncanny, the antiorganic, the life-denying character of them all—each of them was absolutely symmetrical, icily regular in form. They were too regular, as substance adapted to life never was to this degree—the living principle shuddered at this perfect precision, found it deathly, the very marrow of death—Hans Castorp felt he understood now the reason why the builders of antiquity purposely and secretly introduced minute variation from absolute symmetry in their columnar structures."[41]

Weyl's own life and work no less sensitively traced out this interplay between symmetry and life, field and matter, mathematics and physics, reflection and action.

So rich and manifold are Weyl's writings that I have tried to include everything I could while avoiding excessive repetitiveness. I thank Erhard Scholz and Skúli Sigurdsson for their very helpful advice and for the guidance I gained from their own writings about Weyl; Nils Röller, Thomas Ryckman, Brandon Fogel, and Andrew Ayres were most friendly in sharing their thoughts and findings. I am especially grateful to Philip Bartok for giving me essential help with the translations, for which Norman Sieroka also offered invaluable critical guidance and advice; reading his own work on Weyl and corresponding with him was of great help to me. I thank the John Simon Guggenheim Memorial Foundation for its support, as well as Vickie Kearn and her associates at Princeton University Press for their enthusiastic collaboration. Finally, Michael Weyl and Annemarie Weyl Carr were most generous in sharing their recollections.

Not long after making his epochal contributions to quantum theory, Dirac was invited to visit universities across the United States. When he arrived in Madison, Wisconsin, in 1929, a reporter from the local paper interviewed him and learned from Dirac's laconic replies that his favorite thing in America was potatoes, his favorite sport Chinese chess.[42] Then the reporter wanted to ask him something more.

"They tell me that you and Einstein are the only two real sure-enough high-brows and the only ones who can understand each other. I won't ask you if this is straight stuff for I know you are too modest to admit it. But I want to know this—Do you ever run across a fellow that even you can't understand?"

"Yes," says he.

"This will make a great reading for the boys down at the office," says I. "Do you mind releasing to me who he is?"

"Weyl," says he.

The interview came to a sudden end just then, for the doctor pulled out his watch and I dodged and jumped for the door. But he let loose a smile as we parted and I knew that all the time he had been talking to me he was solving some problem that no one else could touch.

But if that fellow Professor Weyl ever lectures in this town again I sure am going to take a try at understanding him. A fellow ought to test his intelligence once in a while.

So should we—and here is Professor Weyl himself, in his own words.

Peter Pesic

1 ▣

Electricity and Gravitation

1921

Modern physics renders it probable that the only fundamental forces in Nature are those which have their origin in gravitation and in the electromagnetic field. After the effects proceeding from the electromagnetic field had been coordinated by Faraday and Maxwell into laws of striking simplicity and clearness, it became necessary to attempt to explain gravitation also on the basis of electromagnetism, or at least to fit it into its proper place in the scheme of electromagnetic laws, in order to arrive at a unification of ideas. This was actually done by H. A. Lorentz, G. Mie, and others, although the success of their work was not wholly convincing.[1] At the present time, however, in virtue of Einstein's general theory of relativity, we understand in principle the nature of gravitation, and the problem is reversed. It is necessary to regard electromagnetic phenomena, as well as gravitation, as an outcome of the geometry of the universe. I believe that this is possible when we liberate the world-geometry (on which Einstein based his theory) from an inherent inconsistency, which is still associated with it as a consequence of our previous Euclidean conceptions.

The great accomplishment of the theory of relativity was that it brought the obvious problem of the *relativity of motion* into harmony with the existence of *inertial forces*. The Galilean law of inertia shows that there is a kind of obligatory guidance in the universe, which constrains a body left to itself to move with a perfectly definite motion, once it has been set in motion in a particular direction in the world. The body does this in virtue of a tendency of persistence, which carries on this direction at each instant "parallel to itself."

At every position P in the universe, this tendency of persistence (the "guiding field") thus determines the infinitesimal parallel displacement of vectors from P to world-points indefinitely near to P. Such a continuum, in which this idea of infinitesimal parallel displacement is determinate, I have designated as "affinely connected."[2] According to the ideas of Galileo and Newton, the "affine connection" of the universe (the difference between straight and curved) is given by its geometrical structure. A vector at any position in the universe determines directly and without ambiguity, at every other position, and by itself (i.e., independently of the material content of the universe), a vector "equal" to itself. According to Einstein, however, the guiding field is a physical reality which is dependent on the state of matter, and manifests itself only infinitesimally (as a tendency of persistence which carries over the vectors from one point to "indefinitely neighboring" ones). The immense success of Einstein's theory is based on the fact that the effects of gravitation also belong to the guiding field, as we should expect a priori from our experience of the equality of gravitational and inertial mass. The planets follow exactly the orbit destined to them by the guiding field; there is no special "gravitational force" necessary, as in Newton's theory, to cause them to deviate from their Galilean orbit. In general, the parallel displacement is "non-integrable," i.e., if we transfer a vector at P along two different paths to a point P' at a finite distance from P, then the vectors, which were coincident at P, arrive at P' in two different end-positions after traveling these two paths.

The "affine connection" is not an original characteristic of the universe, but arises from a more deeply lying condition of things—the "metrical field."[3] There exists an infinitesimal "light cone" at every position P in the world, which separates past and future in the immediate vicinity of the point P. In other words, this light cone separates those world-points which can receive action from P from those from which an "action" can arrive at P. This "cone of light" renders it possible to compare two line-elements at P with each other by measurement; all vectors of equal measure represent one and the same *distance* at P.[4] In addition to the determination of measure at a point P (the "relation of action" of P with its surroundings), we have now the "metrical relation," which determines the congruent transference of an arbitrary distance at P to all points indefinitely near to P.[5]

Just as the point of view of Einstein leads back to that of Galileo and Newton when we assume the transference of vectors by parallel displacement to be integrable, so we fall back on Einstein when the transference of distances by congruent transference is integrable. But this particular assumption does not appear to me to be in the least justified (apart from the progress of the historical development). It appears to me rather as a gross inconsistency. For the "distances," the old point of view of a determination of magnitudes in terms of each other is maintained, this being independent of matter and taking place directly at a distance. This is just as much in conflict with the principle of the *relativity of magnitude* as the point of view of Newton and Galileo is with the principle of the *relativity of motion*. If, in the case in point, we proceed in earnest with the idea of the continuity of action, then "magnitudes of condition" occur in the mathematical description of the world-metrics in just sufficient number and in such a combination as is necessary for the description of the electromagnetic and of the gravitational field. We saw above that, besides inertia (the retention of the vector-direction), gravitation was also included in the guiding field, as a slight variation of this, as a whole, constant inertia. So in the present case, in addition to the force which conserves space- and time-lengths, *electromagnetism* is also included in the metrical relation. Unfortunately, this cannot be made clear so readily as in the case of gravitation. For the phenomena of gravitation are easily obtained from the *Galilean principle*, according to which the world-direction of a mass-point in motion follows at every instant the parallel displacement. Now it is by no means the case that the ponderomotive force of the electromagnetic field should be included in our Galilean law of motion, as well as gravitation, for a charged mass-point does not follow the guiding field. On the contrary, the correct equations of motion are obtained only by the establishment of a definite and concrete law of Nature, which is possible within the framework of the theory, and not from the general principles of the theory.

The form of the law of Nature on which the condition of the metrical field is dependent is limited by our conception of the nature of gravitation and electricity in still greater measure than it is by Einstein's general principle of relativity.[6] When the metrical connection alone is virtually varied, the most simple of the assumptions possible leads exactly to the *theory of Maxwell*. Thus, whereas Einstein's theory of gravitation gave certain inappreciable deviations from the Newtonian theory, such as could be tested by experiment, our interpretation of electricity—one is almost tempted to say unfortunately—results in the complete confirmation of Maxwell's laws. If we supplement Maxwell's "magnitude of action" by the simplest additional term which also allows of the virtual variation of the "relation of action,"[7] we then arrive at

Einstein's laws of the gravitational field, from which, however, there are two small deviations:

1. The *cosmological term* appears, which Einstein appended later to his equations and which results in the spatial closure of the universe. A hypothesis conceived ad hoc by Einstein to explain the generally prevailing equilibrium of masses results here of necessity. Whereas Einstein has to assume a pre-established harmony between the "cosmological constant" which is characteristic for his modified law of gravitation and the total mass fortuitously present in the universe, in our case, where no such constant occurs, the world-mass determines the curvature of the universe in virtue of the laws of equilibrium. Only in this way, it appears to me, is Einstein's cosmology at all possible from a physical point of view.

2. In the case where an electromagnetic field is present, Einstein's cosmological term must be supplemented by an additional term of similar character. This renders the existence of charged material particles possible without requiring an immense mass-horizon as in Einstein's cosmology.

At first the *non-integrability of the transference of distances* aroused much antipathy.[8] Does not this mean that two measuring-rods which coincide at one position in the universe no longer need to coincide in the event of a subsequent encounter? Or that two clocks which set out from one world-position with the same period will possess different periods should they happen to encounter at a subsequent position in space? Such a behavior of "atomic clocks" obviously stands in opposition to the fact that atoms emit spectral lines of a definite frequency, independently of their past history. Neither does a measuring-rod at rest in a static field experience a congruent transference from moment to moment.

What is the cause of this discrepancy between the idea of congruent transfer and the behavior of measuring-rods and clocks? I differentiate between the determination of a magnitude in Nature by "persistence" (*Beharrung*) and by "adjustment" (*Einstellung*).[9] I shall make the difference clear by the following illustration: We can give to the axis of a rotating top any arbitrary direction in space. This arbitrary original direction then determines for all time the direction of the axis of the top when left to itself, by means of a *tendency of persistence* which operates from moment to moment; the axis experiences at every instant a parallel displacement. The exact opposite is the case for a magnetic needle in a magnetic field. Its direction is determined at each instant independently of the condition of the system at other instants by the fact that, in virtue of its constitution, the system *adjusts* itself in an unequivocally determined manner to the field in which it is situated. A priori we have no ground for assuming as integrable a transfer which results purely from the tendency of persistence. Even if that is the case, as, for instance, for the rotation of the top in Euclidean space, we should find that two tops that start out from the same point with the same axial positions and meet again after the lapse of a very long time would show arbitrary deviations of their axial positions, for they can never be completely isolated from every influence. Thus, although, for example, Maxwell's equations demand the conservation equation $de/dt = 0$ for the charge e of electron, we are unable to understand from this fact why an electron, even after an indefinitely long time, always posseses an unaltered charge, and why the same charge is associated with all electrons. This circumstance shows that the charge is not determined by persistence, but by adjustment, and that there can exist only *one* state of equilibrium of the negative electricity, to which the corpuscle adjusts itself afresh at every instant. For the same reason we can conclude the same thing for the spectral lines of atoms. The one thing common to atoms emitting the same frequency is their constitution and not the agreement of their frequencies on the occasion of an encounter in the distant past. Similarly, the length of a measuring-rod is obviously determined by adjustment, for I could not give *this* measuring-rod in *this* field-position any other length arbitrarily (say double or triple length) in place of the length that it now possesses, in the manner in which I can at will predetermine its direction. The theoretical possibility of a determination of length by adjustment is given as a consequence of the *world-curvature*, which arises from the metrical field according to a complicated mathematical law. As a result of its constitution, the measuring-rod assumes a length which possesses this or that value, *in relation to the radius of curvature of the field*. In point of fact, and taking the laws of Nature indicated above as a basis, it can be made plausible that measuring-rods and clocks adjust themselves exactly *in this way*, although this assumption—which, in the neighborhood of large masses, involves the displacement of spectral lines toward the red upheld by Einstein—does not appear anything like so conclusive in our theory as it does in that of Einstein.

2 🔲
Two Letters by Einstein and Weyl
on a Metaphysical Question

1922

[In May 1922 the French physicist Paul Langevin gave three lectures in Zurich on Einstein's relativity theory, the first of which was such a thunderous success that the journalist E. Bovet posed an "easy question" to Langevin: "How can we explain the enthusiasm of the public, which—apart from a few exceptions—surely understood no more of relativity theory than I? Is this pure snobbery? Courtesy to a foreign scholar? Or is it explained through the surmise of a fundamental alteration in our view of the world? Would such a surmise be legitimate? If so, in what sense? Does relativity theory perhaps signify liberation from the mechanistic, materialistic view of the world, under whose pressure our modern culture is breaking up?" Though Langevin did not answer Bovet's personal appeal, Einstein and Weyl did reply.]

Berlin, June 7, 1922
Haberlandstrasse 5

Dear Sir,

Your "Question to Mr. Langevin" provokes me to give an answer. Regarding the general questions that interest you, relativity theory changes nothing at all in the state of affairs because it signifies nothing but an improvement and modification of the basis of the physical-causal world-picture without a change in its fundamental point of view. This is a kind of logical system for representing space-time events in which mental essences (will, feeling, etc.) do not apply directly. To avoid a collision between the various sorts of "realities" that physics and psychology deal with, Spinoza and Fechner respectively founded the theory of psychophysical parallelism, which, quite frankly, completely pleases me.[1] Physics signifies one possible way among others equally justified to put experience in a certain order. The foundations of this system are freely chosen by us, namely from the point of view that at any given time satisfies known facts with a minimum of hypotheses. Thus, this is not a matter of "believing," but rather of free choice from the point of view of logical completeness and adaptability to experience, as indeed is so beautifully shown in the cited passages from Henri Poincaré.[2]

The question "what is the use?" only means something—if it is really supposed to have a clear meaning—when completed by an expression signifying *for whom*, or even better for the satisfaction of whose wish, the thing in question may serve. I really cannot say more than this truism.

A. Einstein

Zurich, July 27, 1922

Dear Sir,

Mr. Bovet's question, to which you invited me to reply, surprised me in two ways. First, that even today, after Western intellectual life has striven for one hundred fifty years to overcome the primitive position of the Enlightenment, that the strict lawfulness of the world of appearances can seem oppressive to the evaluating, willing, and active ego. And second, that Einsteinian dynamics, which only allows the energy and momentum of a body to depend on its velocity a little differently than Newtonian mechanics, is associated with the expectation of an easing of this pressure. Thus, as Mr. Bovet puts the question, one must unhesitatingly answer it in the *negative*; the inexorability of rational mechanics cannot be mitigated through the new view of things. Even a living organism, a rational being, can only put itself in uniform rectilinear motion like any mechanical system by pushing itself away from other bodies, to which it thereby gives an equal and opposite momentum.

Yet it appears to me that physics has no far-reaching meaning for reality, just as formal logic, for example, has no far-reaching meaning in the realm of truth. The foundation of the truth of a judgment lies in the judged thing and not in logic. Every truth in itself is founded with regard to its contents, and (when perceiving) we try to seize this foundation in the depths, through insight, through intuitive reason. Nevertheless, the surface relations, which logic treats, govern the particular truths. But a gagging of the truth-establishing power, of reason, by no means lies therein. In an equal sense, a certain formal constitution of reality is pronounced in the physical laws. These laws will be violated in reality just as little as there are truths not in accord with logic, but these laws do not matter for the essential contents of reality; the ground of reality is not grasped by them. Of course, they do not allow free rein to every whim and caprice, but nothing hinders us from understanding them as surface aspects of a necessity that is "not of this world" and whose reality-grounding power we believe we feel in our moral wills. Likewise, in the domain of knowledge: if, for example, I judge "$2 + 2$ is 4," then I believe that this judgment does not come purely from natural causality in my brain making it so, but instead because the factually existing circumstance $2 + 2 = 4$—thus something not part of the things and forces of reality—has influence on my judgment.

But you do not wish to hear my philosophical point of view about the problem of causality; instead, you want information about whether the new development of physics has brought with it a shift in our understanding of natural causality. This I would like to *affirm*, yet this transformation does not come from relativity theory but from the modern atomic physics of matter. So far as I can judge, most physicists no longer believe in a "Laplacian world-formula," in causality in the sense that, *following simple and rigorously valid mathematical laws, which are investigated once and for all, the state of the cosmos at one moment unequivocally determines its complete past and future.* In physics today, we place atomic matter over against the "ether" or the "field" as the space-time extended medium that transmits the action from material particle to material particle. The sole ultimate constituents of matter are not, like ether, somewhat spatially extended, but each of them simply is inserted into a spatial field-neighborhood from which its field-actions emerge. The "ether"—which one ought not represent to oneself in the image of a substance—joins together all these material individuals into the active whole of a single external world. The cause of the field-states lies in matter; for example, light, which is a field phenomenon, is being excited, is being sent forth from matter. And today it seems as though *rigorous laws underlie the propagation of action in the ether*—with whose arrangement field-physics occupies itself—as though we can only establish *statistical uniformities about how matter causes field-states*; the entire physics of matter is statistical in nature.

According to the view sketched here, matter appears as *an agent* [agens] *that, by virtue of its essence, lies beyond space and time.*[3] This agent composed of innumerable unconnected individuals we call "matter," so far as we consider it as the cause of the actions spreading out in the field by which the individuals weave together a world. According to its inner condition, this agent may just as well be creative life and will as matter. [. . .][4]

H. Weyl

3 ▣
Time Relations in the Cosmos, Proper Time, Lived Time, and Metaphysical Time

1927

1. The possible space-time locations or world points form a four-dimensional continuum in the mathematical sense. This already is said to describe a definite *structure* in this medium of the external world, if one believes that a splitting of the world into an *absolute space* and an *absolute time* has an objective meaning in the sense of saying about two separated, strictly limited space-time events that they take place at the same place (for different times) or the same time (for different places). All world points at equal times form a three-dimensional *stratum*, all world points at the same place form a one-dimensional *fiber*. The structure of the world according to this view also allows itself to be described by saying that the world possesses a fibration and a stratification across the fibers.[1]

If a four-dimensional continuum is referred to by coordinates, then in this way it is described by the four-dimensional number space, the continuum of all quadruplets of numbers. Only in order to make possible a more familiar expression, let the number space, reduced by one dimension, be replaced by an intuitive space endowed with a Cartesian coordinate system. One needs such an arbitrarily chosen mapping in order that the usual geometric-kinematic terms can be applied. Geographical maps are a two-dimensional analogue. With regard to a certain map, one can state that three places on the Earth lie in a straight line, but then one would not wonder whether they do so on another map. Only those relations that remain independent of the chosen mapping under whatever deformation of the picture have objective meaning.—The world-geometrical description is not pictorial, but rather an accurate reproduction of the state of affairs itself, so long as the concept of the continuum is understood in an abstract mathematical sense. The portrayal will be pictorial only if one replaces the number space with the space of intuition.

The stratum passing through a world point O, the *present* for O, separates the *past* and the *future* from one another. The real meaning of this separation is causal, as Leibniz already realized: the abstract time relations in the cosmos should express the *effectual relationship*, the causal structure of the world. With regard to these abstract time relations, though, modern development has led on convincing grounds to an essential correction. The separation into past and future, into that part of the world into which actions can reach out from O and that part out of which actions can reach to O, is brought about not through a "stratum" but through a cone-shaped figure whose apex is at O, the *light cone*. The intuitive comprehension of these relations does not cause the least difficulties if one always inquires about the possible effectual relationship of two events, instead of their time relation. The factually really important distinction between the light cones and the strata lies in the following: if the world point O' lies on the stratum running through O, then the stratum through O' coincides with the stratum through O. If, on the other hand, O' lies on the light cone that originates from the apex O, then the light cone emerging from O' in no way coincides with the light cone originating from O.[2]

2. In order to understand the principal features of the relation between the external world and perceiving consciousness, I simplify my sensual body to a *point-eye*. The point-eye describes a world line. The arrangement and sequence of the points of this world line correspond to the lived "earlier" and "later" of immanent time. A typical case in which everything essential can be seen is the observation of two or more stars. In the

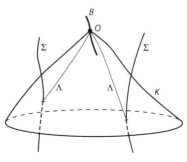

Figure 3.1

figure [3.1] are shown the elements on which the angular distance θ between two stars depends: the world line B of the observer, on it the world point O, at which the observation happens to take place, the backward light cone K emerging from O; further, the two world lines of the stars Σ, which are touched by the cone K each at one point, and the world lines Λ of the two light signals from the stars arriving at O. The magnitude θ will be determined from this by a purely arithmetically describable construction. This magnitude is invariant;

that is, it is of such a kind that an equal magnitude θ results if, after any deformation of the image whatever, θ is calculated once again according to the same procedure on the deformed image. Those angles θ between any two stars in a constellation are decisive for the visual form [*Gestalt*] of the constellation, which cannot be objectively described, but only intuitively experienced. This visual form appears only under the equally objectively ungraspable assumption that *I* am that point-eye. If the angles θ agree with that of a second constellation, both constellations appear in the same visual form; if not, then in different forms.

The immediately experienced is *subjective and absolute*. On the other hand, the *objective* world is necessarily *relative* and may be represented by something definite, numbers or other symbols, only after a coordinate system has been arbitrarily imposed on the world. This pair of opposites, subjective-absolute and objective-relative, seems to me one of the most fundamental epistemological insights one can gather from science. The necessity of the coordinate system goes back to the ultimate epistemological fact, the interpenetration of the *This* (here-now) and the *That*. This interpenetration is the general form of consciousness: only insofar as continuous extension and continuous quality coincide does something exist. This double nature of that which is real has the consequence that we can only draw up a theoretical picture of that which exists *against the background of the Possible*. Thus, in particular, the extensive four-dimensional medium of the world is the field of possible coincidences. The coordinate system becomes inevitable because we must grasp the structure of the world that cannot be read off from real events, but rather only read off from the abundance of those events that are in possible compliance with natural law.

According to Einstein, the world has an *objective determination of measure* according to which the parts of the world line of a body can be measured against each other (*proper time*). In order to make provisional contact with experience, Einstein defines the proper time through the readings of a *clock* carried with the motion of a body. In truth, though, the behavior of the clock under the influence of the metrical field is derived from its own material constitution and from the causal laws (*Wirkungsgesetzen*); this behavior cannot be prejudiced through a definition.—The lived time of consciousness also has in itself a vague measurement, beside the ordering of earlier and later; there is an *immediate estimate of lived time*. For it, likewise, proper time is surely the physical basis, but proper time depends, no less than the timekeeping of a clock, on many accidental circumstances that the psychologist has sought to investigate accurately.

3. *Lived time* is an enduring *Now*, filled with changing contents. But every momentary phase of the lived is changed from the mode of Now-being to the mode of Just-having-been; this subsidence into pastness, captured in retention, like the protention of the future, rising up into the present, belongs to immediate time-life as well. On the other hand, reminiscence is an act of bringing to mind elements of consciousness that, as already-lived [*gelebte*], have disappeared; here already are constituted objective structures whose sensual content is held as identical, independently of the to-be-lived. Time is thereby ordered, through the connection of earlier and later, into a one-dimensional continuum of time points to which the lived contents are bound. Consciousness glides along this time line and awakens to life one point after another to the life of the Now, the immediate present. The world extended in a four-dimensional medium simply *is*, does not *happen*. Only in the look of a consciousness creeping upwards along the world line of its body does a section of this world come to life and pass by as a picture, grasped as spatial and as being in temporal transformation.

In this portrayal, *metaphysical time* comes forward as the connecting link between objective time, the mathematical ordering scheme of the points on the world line of the I-body, and time lived by me. Of course the objective world, expressible only in mathematical symbols, can be won only from the given in experience—through abstraction, objectification, totalization, projection on the horizon of the possible; if one severs this connection, there remains a pure play of signs without "meaning." But even if in this way what is given by consciousness is epistemologically prior, still reason cannot help positing the objective world as that which is prior with respect to the grounds of Being. *Metaphysics* is the attempt to accomplish this reversal. Relativity theory has thus a metaphysical meaning insofar as it has taught, or at least has confirmed from the side of physics, that the far-reaching modification of Being through time we express through the word "Now-being" will not imprint the world but will imprint the monads.[3] The question about the possibility of metaphysics will not be touched here; in this connection, the above-stressed opposition between subjective-absolute and objective-relative entails a clear warning, inasmuch as the endeavor of metaphysics is directed toward objective-absolute Being.

The point-nature of the Now within the time continuum raises a certain difficulty within the conception of metaphysical time, for within a continuum, a point, without the neighborhood through which it is bound to the whole continuum, is not capable of existence. A point in a continuum is not an element of a set, but rather an ideal boundary of continuous partitions. And yet here the present is presumed to enter into consciousness in strictly point-like fashion and to expire again at that same point in time. In my opinion, here the facts of atomic theory show a way out. All the physical characteristics of the ultimate elementary particles of matter, particularly of electrons, can be read off from the neighboring *field*; the application of geometrical, mechanical, physical concepts to the electron itself and its extension seem to be without meaning. Accordingly, one would like to treat material particles as something otherworldly, not taking up extension. Such a particle is itself not spatial, but only lies within a spatial neighborhood, from which its actions originate. Using a half-pictorial turn of phrase, one might say that the world-continuum grows out of a purely fictional seamless continuum through cutting individual world tubes (*Weltröhren*) from within, which arrange the world line of the various material particles heretofore appearing in our interpretation.[4] Yet the inside of the tubes, including their bounding shell, no longer belongs to the world, but is a seam that, like the infinitely distant, is unreachable from within the field. Then there is no pointlike Now and also no exact earlier and later. Roughly speaking, then things act as if the life-point bound into the body, which awakens the objective world into existence for consciousness, possesses not only a diffuse spatial extension but also a diffuse temporal extension. The immediate present is not entirely abrupt; there is always a small halo, quickly fading toward the past and toward the future, along with the self-shining light of immediacy.

4

The Open World: Three Lectures
on the Metaphysical Implications of Science

1932

Preface

One common thought holds together the following three lectures: Modern science, insofar as I am familiar with it through my own scientific work, mathematics and physics make the world appear more and more as an open one, as a world not closed but pointing beyond itself. Or, as Franz Werfel expresses it in pregnant wording in one of his poems,

"Diese Welt ist nicht die Welt allein."

Science finds itself compelled, at once by the epistemological, the physical and the constructive-mathematical aspect of its own methods and results, to recognize this situation. It remains to be added that science can do no more than show us this open horizon; we must not by including the transcendental sphere attempt to establish anew a closed (though more comprehensive) world.

I am grateful to Yale University for affording me an opportunity in these Terry Lectures of expressing this conviction by a description of the methodology of mathematics and physics. The lectures were originally written out in German. I do not want to omit acknowledging my indebtedness to my friend, Dr. Lulu Hofmann of Columbia University, New York, for the devoted assistance which she has rendered me in the translation of my manuscripts into English on this as well as on similar previous occasions.

H. W.
Yale University
April 15, 1931

I. God and the Universe

A mathematician steps before you, speaks about metaphysics, and does not hesitate to use the name of God. That is an unusual practice nowadays. The mathematician, according to the ideas of the modern public, is occupied with very dry and special problems, he carries out increasingly complicated calculations and more and more intricate geometrical constructions, but he has nothing to do with those decisions in spiritual matters which are really essential for man. In other times this was different. Pythagoras, whose figure almost merges into the darkness of mythology, by his fundamental doctrine that the essence of things dwells in numbers, became at the same time the head of a mathematical school and the founder of a religion. Plato's profoundest metaphysical doctrine, his doctrine of ideas, was clad in mathematical garb when he expounded it in rigorous form; it was a doctrine of ideal numbers, through which the mind was to apprehend the structural composition of the world. The spatial figures and relations investigated by geometry—half notional category, half sense perception—were to him the mediators between the phenomenon and the idea. He refused admission to the academy to those who were not trained in mathematics. To Plato, the mathematical lawfulness and harmony of nature appeared as a divine mind-soul. The following words are from the twelfth book of the *Laws*:

> There are two things which lead men to believe in the Gods: one is our knowledge about the soul, as being the most ancient and divine of all things; and the other is our knowledge concerning the regularity of the motion of the stars and all the other bodies.
>
> The present opinion is just the opposite of what once prevailed among men, that the sun and the stars are without soul. Even in those days men wondered about them, and that which is now ascertained was then conjectured by some who had a more exact knowledge of them—that if they had been things without soul, and had no mind (*nous*) could never have moved with numerical exactness so wonderful; and even at that time some ventured to hazard the conjecture that mind was the orderer of the universe. But these same persons again mistaking the nature of the soul, which they conceived to be younger and not older than the body, once more overturned the world,

35

or rather, I should say, themselves; for the bodies which they saw moving in heaven all appeared to be full of stones and earth and many other lifeless substances, and to these they assigned the causes of all things. Such studies gave rise to much atheism and perplexity, and the poets took occasion to be abusive. . . . But now, as I said, the case is reversed.

No man can be a true worshipper of the Gods who does not know these two principles—that the soul is the oldest of all things which are born, and is immortal and rules over all bodies; moreover, as I have now said several times, he who has not contemplated the mind of nature which is said to exist in the stars, and gone through the previous training, and seen the connection of music with these things, and harmonized them all with laws and institutions, is not able to give a reason for such things as have a reason.

The cosmology of Aristotle, with its distinction between the terrestrial sublunar domain and the heavenly sphere set into revolution by the "unmoved primal mover," in combination with the Ptolemaic world system which places the earth in the center of the universe, forms the fixed frame into which the medieval church built its dogma of God, Savior, angels, man, and Satan. Dante's *Divina Commedia* is not only a poem of great visionary power, but it contains a bold theological and geometrical construction of the cosmos, by means of which Christian philosophy adapts Aristotle's cosmology to its own use. While the Aristotelian universe is enclosed by a sphere, the crystal sphere, beyond which there is no further space, Dante lets the radii emanating from the center of the earth, the seat of Satan, converge toward an opposite pole, the source of divine force, much as on the sphere the circles of longitude radiating from the south pole reunite at the north pole. The force of the personal God must radiate from a center, it cannot embrace the world sphere reposing in spatial quiescence like the "unmoved primal mover" of Aristotle. To sense perception, of course, Aristotle's description remains valid. The innermost circles, which surround the divine source of light most closely, by being most heavily charged with divine force become spatially most comprehensive and encompass the more removed circles. In modern mathematical language we would say that Dante propounds a doctrine which in our days has been reestablished by Einstein for entirely different reasons, the doctrine, namely, that three-dimensional space is closed, after the manner of a two-dimensional spherical surface; but from the pole of divine force there radiates a metric field of such a nature that spatial measurement leads to the conditions described by Aristotle.[1]

The Aristotelian world concept was shaken by Copernicus, who recognized the relativity of motion. How could this knowledge, epistemological and mathematical in character and of such complexity that the precise formulation of it even now surpasses the average man's capacity for abstraction—in spite of its being taught, in a somewhat coarse and dogmatic form, of course, in our schools—how could this insight inaugurate a new era in natural philosophy? Only through its coalescence with a certain religious attitude of man toward the universe; for it deprived the earth, the dwelling place of mankind, of its absolute prerogative. The act of redemption by the Son of God, crucifixion and resurrection are no longer the unique cardinal point in the history of the world, but a hasty performance in a little corner of the universe repeating itself from star to star: this blasphemy displays perhaps in the most pregnant manner the precarious aspect which a theory removing the earth from the center of the world bears for religion. In this respect Giordano Bruno drew the conclusions with vehement enthusiasm. Aristotle, Ptolemy, and the ecclesiastical dogma were to him the "three-headed scholastic beast" with which he struggled throughout his life of unrest. To him there lay a mighty liberation in the transition from Aristotle's world, enclosed in the crystal sphere and ordered hierarchically according to strictly distinguished forms of being, to the indifferent expanse of infinite Euclidian space which is everywhere of the same constitution and everywhere filled with stars—the concept which is the foundation of the new natural philosophy. In his *Schriften zur Weltanschauung und Analyse des Menschen seit Renaissance und Reformation*, Dilthey says: "The foundation of the more recent European pantheism is the recognition of the homogeneity and the continuous connection of all parts of the universe." Nicolaus Cusanus [Nicholas of Cusa] and Giordano Bruno are the first heralds of the new conception. Like Pythagoras before him, Bruno considers himself the proclaimer of a "Holy Religion" on the ground of a new mathematical cognition. To him the change from the anthropocentric view supported by sense appearance to the cosmocentric one acquired by astronomy is only one part of the great revolution effected in the human mind by the new Copernican epoch. There corresponds to it an equally deep and thorough revolution in the religious and moral domain. Sensual consciousness has its center in the preservation of the physical existence which is confined between birth and death. With the emancipation from sense appearance as a result of astronomical discoveries and their philosophical utilization, there is connected the elevation of man to the love of God and of his cosmic manifestations. Not until now do we perceive the true perfection of the universe which springs from the relation of its parts to the whole, and thereby relinquish the undue demands made of this divine order, demands which have their source in the desires of the individual to perpetuate his own existence.

The ideas of Bruno, which propagate themselves in their influence on Spinoza and Shaftesbury, lead to a justified, thoroughgoing, valuable, and promising transformation in the religious attitude of occidental Christianity. Religious belief will always center about two issues, the one cosmic in character, emphasizing human dependence on and relationship to the universe; the other personal, involving moral dignity, autonomy and individual responsibility. In both of these respects, however, a change and advance takes place which is demanded by the progress of culture. This seems to have been the conviction also of the founder of these lectures. But the more modern science, especially physics and mathematics, strives to recognize nature as it is in itself or as it comes from God, the more it has to depart from the human, all too human ideas with which we respond to our practical surroundings in the natural attitude of our existence of strife and action. And the more strange and incomprehensible it must necessarily become to those who cannot devote their entire time and energy to the development and readjustment of their theoretical thinking; herein lies the actual and inevitable tragedy of our culture. For the philosophical and metaphysical import of science has not declined but rather grown through its estrangement from the naive world of human conceptions.

So far I have been speaking of astronomical research and cosmological speculation, with reference to the manner in which our conception of God and divine action in nature is formed and transformed together with such speculation. I shall return to this point later a little more systematically. But quite aside from the fact that mathematics is the necessary instrument of natural science, purely mathematical inquiry in itself, according to the conviction of many great thinkers, by its special character, its certainty and stringency, lifts the human mind into closer proximity with the divine than is attainable through any other medium. *Mathematics is the science of the infinite*, its goal the symbolic comprehension of the infinite with human, that is finite, means. It is the great achievement of the Greeks to have made the contrast between the finite and the infinite fruitful for the cognition of reality. The intuitive feeling for, the quiet unquestioning acceptance of the infinite, is peculiar to the Orient;

but it remains merely an abstract consciousness, which is indifferent to the concrete manifold of reality and leaves it unformed, unpenetrated. Coming from the Orient, the religious intuition of the infinite, the *apeiron*, takes hold of the Greek soul in the Dionysiac-Orphic epoch which precedes the Persian wars.[2] Also in this respect the Persian wars mark the separation of the Occident from the Orient. This tension between the finite and the infinite and its conciliation now become the driving motive of Greek investigation; but every synthesis, when it has hardly been accomplished, causes the old contrast to break through anew and in a deepened sense. In this way it determines the history of theoretical cognition to our day.

The connection between the mathematics of the infinite and the perception of God was pursued most fervently by Nicholas of Cusa, the thinker who as early as the middle of the fifteenth century, sometimes impetuously, sometimes full of prophetic vision, intoned the new melody of thought which with Leonardo, Bruno, Kepler, and Descartes gradually swells into a triumphant symphony. He recognizes that the scholastic form of thinking, Aristotelian logic, which is based on the theorem of the excluded third, cannot, as essentially a logic of the finite, attain the end for which scholasticism employed it: to think the absolute, the infinite. It must always and of necessity break down where the perception of the infinite is in question. Thereby every kind of "rational" theology is rejected, and "mystic" theology takes its place. But Cusanus is beyond the traditional notion of logic as well as the traditional notion of mysticism; for with the same determination with which he denies the cognition of the infinite through the logic of the finite, he denies the possibility of its apprehension through mere feeling. The true love of God is *amor Dei intellectualis* [the intellectual love of God]. And to describe the nature and the aim of the intellectual act through which the divine reveals itself to us, Cusanus does not refer to the mystic form of passive contemplation, but rather to mathematics and its symbolic method. *"Nihil veri habemus in nostra scientia nisi nostram mathematicam."* [We have nothing true in our science beside our mathematics.] On the one side stands God as the infinite in perfection, on the other side man in his finiteness; but the Faustian urge driving him toward the infinite, his unwillingness to abide with anything once given and attained, is no fault and no *hybris* but evidence of his divine destination. This urge finds its simplest expression in the sequence of numbers, which can be driven beyond any place by repeated addition of the one. We witness here a strange occurrence, unique in the history of philosophy: the exactness of mathematics is sought not for its own sake, nor as a basis for an explanation of nature, but to serve as a foundation for a more profound conception of God. Cusanus is one of the epoch-making minds both in theology and mathematics. All wise men, all the most divine and holy teachers, so his work *De docta ignorantia* [On Learned Ignorance] sets forth, agree that every visible thing is an image of the invisible, which to us is imperceptible except in a mirror and in enigmas. But even if the spiritual in itself remains inaccessible to us, and even if it can never be perceived by us except in images, or symbols, yet we must at least postulate that the symbols themselves contain nothing doubtful or hazy: the symbols must be endowed with the determinateness and the systematic coherence that is possible only on the basis of mathematics. From here the way leads to Leonardo, Kepler, and Galileo who, after two thousand years of mere description of nature, initiate an actual analysis, a theoretical construction of nature with symbolical mathematical means. With regard to the essence of mathematical knowledge, considered as a symbolical *mathesis universalis* [universal knowledge], Cusanus had visions, and expressed ideas, which do not recur in more determinate form until the days of Leibniz; visions, indeed, of which we seem to be acquiring full understanding only at present in the latest attempts to master the antinomies of the infinite by purely symbolical mathematics. This subject will be dealt with in the third lecture.

For speculative metaphysicists, according to Galileo's *Saggiatore*, philosophy is like a book, a product of pure imagination, such as the *Iliad* or *Orlando Furioso*, in which it is of little importance whether what is said is true.

> But that is not so; for philosophy is written in the great book of nature which is continually open before our eyes, but which no one can read unless he has mastered the code in which it is composed, that is, the mathematical figures and the necessary relations between them.

The ideality of mathematics lifts the human mind to its most sublime height and perfection: the barriers erected between nature and the mind by medieval thought break down before it, in a certain sense even the barriers between the human and the divine intellect. I once more quote Galileo:

> It is true that the divine intellect cognizes the mathematical truths in infinitely greater plenitude than does our own (for it knows them all), but of the few that the human intellect may grasp, I believe that their cognition equals that of the divine intellect as regards objective

certainty, since man attains the insight into their necessity, beyond which there can be no higher degree of certainty.

And Kepler: "The science of space is unique and eternal and is reflected out of the spirit of God. The fact that man may partake of it is one of the reasons why man is called the image of God."

After this historical introduction I turn to the question which is to be the primary subject of this lecture: How does the divine manifest itself in nature? As far as I see, this question has been answered chiefly in two ways in the history of human thought. Both answers are forceful, but they are essentially different. The first is more primitive and more objective: the ether is the omnipresence of God in things. The second is more advanced and more formal: the mathematical lawfulness of nature is the revelation of divine reason.

The significance of the ether concept can only be understood in connection with the fundamental ideas of the theory of relativity. Space, the manifold of space points, is a three-dimensional continuum. This manifold is, to begin with, amorphous, without structure; in this condition nothing about it would be changed if I subjected it to some continuous deformation such as one might apply to a mass of clay. Only statements concerning the distinctness or coincidence of points and the continuous connection of point configurations can be made at this stage. But beyond that, space is endowed with structure; this becomes apparent in the fact, among others, that we are able to distinguish the straight lines from the curved ones. A point and a direction assigned to this point uniquely determine a line which passes through it and is of the type we characterize by the adjective straight or geodesic.[3] At earlier times it was believed that among the straight lines the class of verticals was in itself distinguished, that space was designed about the direction from above to below as the original one. We know today that this can be the case only in the gravitational field, where the direction of gravity is distinguished as the one which freely falling bodies follow, but that this direction is determined physically and varies with the physical conditions. The direction from above to below is different in Calcutta from what it is in New Haven, and the angle which these directions form with each other would change if the distribution of mass on the earth were changed, for example, by the folding up of a high range of mountains in the neighborhood of Calcutta. This example seems appropriate to make clear the difference between a rigid geometric structure that cannot be influenced by material forces, like the so-called projective structure which makes possible the distinction between straight and curved, and a structure depending on material influences and changeable with them, as exists, for example, in the directional field of gravitation.

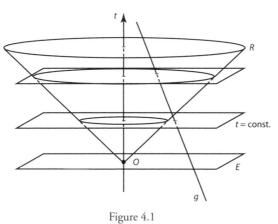

Figure 4.1

In reference to natural phenomena one cannot consider space separately, but one has to connect it with time. By saying "here-now" we fix a space-time point or world point by direct specification. We may mark it by the momentary flash of a spark of light. The possible world points or places of localization in space and time form a four-dimensional continuum. A small body describes a world line, the one-dimensional continuum of the world points which it gradually passes through in the course of its history. It has a meaning directly evident to our intuition only to say of two events that they occur at the same space-time position or in immediate space-time proximity. If one believes in a decomposition of the world into an absolute space and an absolute time, so that it has a meaning to say of two distinct events, closely limited in space-time, that they occur at the same place but at different times, or at the same time but at different places, then one is already assigning a definite structure to the four-dimensional extensive medium of the external world. All simultaneous world points form a three-dimensional stratum, all equipositional world points a one-dimensional fiber. The structure of the world, according to this point of view, can thus be described by stating that it is composed of a stratification traversed by fibers. As long as one cannot refer to a structure of this kind, it is permissible to speak of rest or motion of a body K only with reference to a medium which continuously fills space, or to a body of reference in which K is embedded or on which K lies. In everyday life, the "firm well-founded earth," for good reasons, provides such a body of reference.[4] But who tells us that the earth stands still, or rather, what do we mean by it? The belief that simultaneity exists in the world is originally based on the fact that every person places the events which he perceives in the moment of their perception. But this naive belief lost its ground long ago through the discovery of the finite velocity of the propagation of light.

The theory of relativity clearly recognized that the structure of the world is not a stratification and fibration according to simultaneity and identity of position. It points out: (1) Not rest but uniform translation is an intrinsically distinguished class of motion,

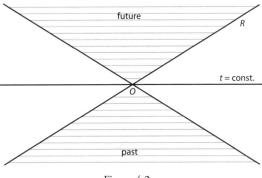

Figure 4.2

uniform translation being the state of motion of a body left to itself and not deviated through the action of any external forces. The world line of such a body is uniquely determined by the starting point and the initial direction of its motion in the world; the "projective" structure which thus manifests itself is called by the physicist the inertial guiding field. The so-called law of inertia, according to which a body that is left to itself moves through space along a straight line with constant velocity (into the discussion of which, however, I cannot enter here), conceives the inertial structure as a rigid geometric entity. (2) The strata of simultaneity are replaced by a causal structure: from every world point O there extends into the world a three-dimensional cone-shaped surface which in the manner evident from the figure determines a region of the past and one of the future from O. If I am now at O, the events on which my actions at O can still be of influence, i.e., those world points which can be reached by an action at O, belong to the future, while those events which exercise an effect upon the events at O are localized in the past. Past events, then, are those of which I, at O, can somehow receive intelligence through direct perception and tradition or recollection based upon such perception; for every perception and every mode of information is a physical transference of action. But between past and future there extends an intermediate region with which I am causally connected neither actively nor passively at this moment. In the old theory past and future touch without intervening space in the stratum of the present, the ensemble of world points simultaneous with O. The abstract time relations must be replaced everywhere by concrete causal connections. The actual motion of a body results from a conflict between the inertial guidance and the deviating forces. The frequently quoted example of a train collision serves as a clear illustration of how the conflict between inertia and the molecular forces of elasticity tears asunder the parts of the train. Thus we see that the structure—no matter how it may have to be described exactly—is of most decisive influence on the course of events. It is the physicist's problem to ascertain it from the physical effects which it produces.

After these general remarks concerning the relativity problem, I shall outline to you the history of the ether concept briefly as follows. In the philosophy of the Stoa, the ether occurs first as the divine fire spread out through the world, as the substratum of the divine creative forces at action in the world. One can read about the Stoic ether theory, for example, in the second and third books of Cicero's work *De natura deorum* (On the Nature of the Gods).[5] In the transition period to which Giordano Bruno belongs, this idea mingles with the atomistic world concept which, after its formulation in antiquity by Democritus, was taken over and developed by the Epicureans. For Bruno the ether is an extended real entity which permeates all bodies but is itself without limits. Here natural science takes hold of the notion and finds in this hypothetical medium an appropriate carrier for the propagation of the natural forces, especially of light, where the ordinary bodies which are perceptible to our senses by their resistance do not suffice. With Huygens, and also with Euler, we encounter the light ether as a substance whose state is determined by its density and its velocity, and which is spread out continuously. Since it as a whole is at rest and is excited to perform only tiny vibrations, it could at the same time serve to provide physical reality (hypothetical indeed) for Newton's metaphysical notion of absolute space. But history took the opposite course. When in the nineteenth century optical phenomena reveal themselves as part of the larger class of electrodynamic ones, and the notion of an electromagnetic field which no longer requires a substantial substratum is developed by Faraday and Maxwell, the ether divests itself of its physical character, and there remains absolute space, a structural element, which is no longer affected by matter as was the light ether. This second stage was anticipated in Newton's natural philosophy. At the beginning of his *Principia*, Newton proclaims with perfect clarity absolute space and absolute time as the entities that are a priori at the bottom of all laws of nature. If one asks how Newton could embrace this dogma although he adopts the empirical program of deriving the actual run of the strata and fibers of the world according to space and time from their effects on the observable events, the answer to my mind lies chiefly in his theology, the theology of

Henry More. Space is to him *sensorium Dei*, the divine omnipresence in all things. Therefore the structure of space behaves with regard to things as one would naturally imagine the behavior of an absolute God toward the world: the world is submitted to his action, but he himself is beyond the influence of any action from the world. Certainly Newton's view of the world thereby acquires a somewhat rigid and scholastic character. In his doctrine about the center of the world and the position of the sun among the fixed stars, for example, he is considerably more Aristotelian and less modern than Giordano Bruno, who precedes him by more than a century. Nevertheless we must admit that the transition from the Stoic ether as a deified potency of nature which is drawn into the play of the natural forces, to the geometrically rigid absolute space represents an advance of conception which is entirely in line with the transition from a mythical religion of nature to the transcendent God of Christianity.

In the third stage of the development it becomes manifest that the space-time structure is described incorrectly by the notion of absolute space; that not the state of rest but of uniform translation is an intrinsically distinguished class of motion. This recognition at the same time completely puts an end to the substantial ether. Newton himself was able to pass from uniform translation to rest only by a strange scholastic trick which shows up most peculiarly in the disposition of the *Principia* which is otherwise so rigorous. Finally, in the fourth place, the general theory of relativity permits this world structure in its inertial as well as its causal aspect again to become a physical entity which yields to the forces of matter; thus in a certain sense the circle closes, even though the quantities of condition characterizing the state of the ether have now become entirely different from those in the beginning of the development, when it appeared on the scene as a substantial medium. For this, indeed, is the fundamental physical thought underlying Einstein's general theory of relativity, that that which produces as powerful and real effects as does this structure cannot be a rigid geometric constitution of the world fixed once and for all, but is something real which not only acts upon matter but also reacts under its influence. In the dualism between inertial guidance and force, as Einstein further recognized, gravitation belongs on the side of inertia; the suspected variability of the inertial field and its dependence on matter manifest themselves in the phenomena of gravitation. As seen from Newton's philosophy, the theory of relativity thus deprives space of its divine character. We now distinguish between the amorphous continuum and its structure: the first retains its a priori character, but becomes the counterpart of pure consciousness, while the structural field is completely given over to the real world and its play of forces. As a real entity it is denoted by Einstein for good historic reasons by the old name of the ether.

The reason why the dependence of the ether upon matter was so hard to discern lies in the extreme predominance of the ether in its interaction with matter—nor does the Einstein theory deny its overruling power. If it is not a god, it is certainly a superhuman giant. One can estimate the proportion of power to be as $10^{20} : 1$ in a sense that can be exactly specified mathematically on the ground of the laws of nature.[6]

If the ether were not disturbed by matter it would abide in the condition of rest, or, speaking mathematically and more precisely, of homogeneity. In contrast with the "spirit of unrest" that dwells in matter, "the breast of the earth and of man," in Hölderlin's language, the ether represents the lofty, hardly disturbed quiescence of the universe.[7] Being of the same essence as, and in principle on the same plane with matter and its forces, the ether is not the divine in nature; still we encounter in it a power which to us human beings—who are able to produce action only through the agent of matter—is strange, overwhelming, soothing, and before which a feeling of deep reverence is well appropriate. In this spirit the great German romantic poet Hölderlin, as late as the beginning of the nineteenth century, devoted to "Father Ether" powerful cosmic songs. And to present day natural philosophy it is still a profound enigma, to my mind *the* deepest mystery with which it is confronted, how this powerful predominance of the ether in its interaction with matter is to be understood. What I primarily wanted to make clear to you by my exposition is how at this point age-old religious and metaphysical ideas and questions are intimately connected with the ultimate problems of actual science.

But no matter how exalted the natural power is which modern physics denotes as gravitational and inertial ether, to us who are Christians and not heathens the ether does not reveal the face of the divine ultimate essence of things. Therefore I shall now follow the line of thought which man has advanced as a second fundamental answer to the question as to the finger of God in nature: the world is not a chaos, but a *cosmos harmoniously ordered by inviolable mathematical laws*. This idea hardly has a history. We suddenly encounter it in completed form with the Pythagoreans, and from them it passes over into Platonic philosophy. Historically I should like to trace it to two sources that lie far apart. First to the age-old number-mysticism and number-magic, an inheritance of mankind from prehistoric times with which the general lawfulness of nature, in the philosophy of Pythagoras, the founder of the school, still appears to have been closely bound up. The law of musical harmony, according to which the harmonic tones are produced by a division of the chord in integral proportions, ranks as the primal law. Following this example, one sought to reduce to integral proportions also the evident regularity in the course of the stars, especially the planetary orbs and their periods of revolution, and called this the harmony of the spheres. Kepler ardently devoted himself to such studies and finally wrought from them his three famous laws of planetary motion, which inaugurated the transition to a more profound conception of the mathematical harmony in the laws of nature. The second source, the anthropomorphic origin of the idea of lawful determination in the cosmos lies in the idea of fate. The acting ego in its existence of strife not only encounters the thou, the fellow man and the fellow animal, but also the resistance of a form of being essentially different from his own and towering gigantically above him: earth and ocean, fire, storm and stars. Their manifestations were first considered and designated by language as acts, as the manifestations of an acting being; for example, we say "The sun shines." As soon as this primitive animism is overcome, as soon as the essential difference of external events as contrasted with acts of the ego born out of a cloudy mixture of insight and urge is recognized and stressed, it appears as fate, *moira*, "Ananke [Necessity], the compelled compulsion" (as it reads in Spitteler's epos, *The Olympic Spring*), with the attribute of unyielding, blind necessity, which is self-sufficient and in no relation to any meaning.[8] But dark ordinance by fate and dark number-magic are overcome in Greek thought by the glorious, lucid idea of mathematical lawfulness governing the world. Without having relation to any meaning as do the acts of the ego, the external world is thus nevertheless filled with the light of spirit, of reason. We know in how wonderful a manner and to what extent this idea about the structure of the external world has stood the test, at first with regard to the motion of the stars, and later also with regard to the confused processes on the earth— since with Kepler and Galileo the course of the world was reduced by an actual analysis of nature to the changes in space-time of measurable quantities of condition. These men succeeded in the most difficult of all advances of the human mind—that of subjecting speculative imagination and a priori mathematical construction to reality and to experience as it is systematically questioned in the experiment.

In a law of nature, as we shall later establish more precisely, simplicity is essential. "Nature loves simplicity and unity," we read in Kepler. The closely allied category of perfection played a great part in Aristotelian philosophy, not only as a methodical but as an explanatory principle. Thus, according to Aristotle, the indestructibility and immutability of the heavenly bodies asserted by him arise from their perfect spherical form. In the polemics which Galileo directed against this conception in his dialogue on the two principal world systems, we feel most keenly the radical change in the interpretation of nature that was made by Galileo. He recognizes the idea of perfection, but he no longer seeks it in fixed forms and individual things—here he praises mutability: how the plant developing into a flower is something incomparably more glorious than the crystal perfection of the bodies in the Aristotelian world, removed from all changes. He seeks it in the dynamic connections and their lawfulness, and finds perfection no longer an objective ultimate constituent of physical properties, but rather a heuristic principle and a creed conducive to research. In the evolution of Kepler's ideas this change also takes place. At the beginning he still adheres to static principles, he attempts to discover the harmony of the planetary system in the scheme of regular bodies. Only gradually and laboriously does he struggle through to a more dynamic conception of the world. "Kepler, Galileo, Bruno," says Dilthey, "share with the antique Pythagoreans the belief in a cosmos ordered according to highest and most perfect rational mathematical laws, and in divine reason as the origin of the rational in nature, to which at the same time human reason is related." On the long road of experience throughout the following centuries, this belief has always found new and surprising partial fulfillments in physics, the longer the more, the most beautiful perhaps in the Maxwell theory of the electromagnetic field. No general notion concerning the essence of the external world can be placed parallel to this one in depth and solidity; although we must admit that nature has again and again proved superior to the human mind, and has forced it to abandon a preliminary conclusion, at times attempted even in a universal world law, for the sake of a deeper harmony.

It was natural for man to attribute the cause for the lawfulness of the world to the reign of souls endowed with reason. I may remind you of the words of Plato quoted at the beginning of this lecture. Kepler finds it hard to understand the obedience of the planets to his second law, which sets the velocity of the planet in functional dependence on its distance from the sun, except by assuming a planetary soul which receives within itself the image of the sun in its changing magnitude. Much more tenaciously than such a psychical interpretation has a mechanical and mechanistic interpretation of the laws of nature tried to assert and maintain itself in physics. Think of Ptolemy's mechanism of wheels, think also of the multitudinous attempts to explain gravitation and all physical phenomena by the impact of hard particles. But physics has had to free itself more and more both from mechanical and psychical interpretations; in atomistic physics, this appears to have taken place only in the latest phases of development of the quantum theory. In an address given at the monument which his birthplace, Weil der Stadt, dedicated to Kepler, Eddington recently spoke of the fact that in Kepler's conception of the world, the music of the spheres was not drowned by the roar of machinery and that herein lies a deep relationship between his astronomical thinking and the development of modern physics. The harmony of the universe is neither mechanical nor psychical, it is mathematical and divine.

The Pythagoreans, and following them Plato, conceived the mathematical regularity of the cosmos merely as an order which does not bind nature or the divine in any sense other than that in which reason, for example, as an agency realizing truth, is bound by the formal logical laws. But later Stoa and Christianity, with their increased accentuation of the value of the individual soul, again amalgamated the idea of the cosmos with that of fate, and in the regularity of nature stressed less the order than the necessity and determination which govern relentlessly the course of all events, including human acts. With Hobbes this results in the modern positivistic determinism of which I shall have to give a more detailed discussion in the second lecture on causality.

I shall conclude this lecture with an epistemological consideration.

The beginning of all philosophical thought is the realization that the perceptual world is but an image, a vision, a phenomenon of our consciousness; our consciousness does not directly grasp a transcendental real world which is as it appears. The tension between subject and object is no doubt reflected in our conscious acts, for example, in sense perceptions. Nevertheless, from the purely epistemological point of view, no objection can be made to phenomenalism which would like to limit science to the description of what is "immediately given to consciousness." The postulation of the real ego, of the thou and of the world, is a metaphysical matter, not judgment, but an act of acknowledgment and belief. But this belief is after all the soul of all knowledge. It was an error of idealism to assume that the phenomena of consciousness guarantee the reality of the ego in an essentially different and somehow more certain manner than the reality of the external world; in the transition from consciousness to reality the ego, the thou and the world rise into existence indissolubly connected and, as it were, at one stroke.

But the one-sided metaphysical standpoint of realism is equally wrong. Viewed from it, egohood remains a problem. Leibniz thought he had solved the conflict between human freedom and divine predestination by letting God (for sufficient reasons) assign existence to certain of the infinitely many possibilities, for example to the beings Judas and Peter, whose substantial nature then determines their entire fate. The solution may be sufficient objectively, but it breaks down before the desperate outcry of Judas: "Why did I have to be Judas?" The impossibility of an objective formulation of the question is evident; therefore no answer in the form of an objective cognition can ensue. Only redemption of his soul can be the answer. Knowledge is unable to harmonize the luminous ego (the highest, indeed the only forum of all cognition, truth, and responsibility) which here asks in despair for an answer, with the dark, erring human being that is cast out into an individual fate. Furthermore, postulating an external world does not guarantee that it shall constitute itself out of the phenomena according to the cognitive work of reason as it establishes consistency. For this to take place it is necessary that the world be governed throughout by simple elementary laws. Thus the mere postulation of the external world does not really explain what it was supposed to explain, namely, the fact that I, as a perceiving and acting being, find myself placed in such a world; the question of its reality is inseparably connected with the question of the reason for its lawful mathematical harmony. But this ultimate foundation for the ratio governing the world, we can find only in God; it is one side of the Divine Being. Thus the ultimate answer lies beyond all knowledge, in God alone; flowing down from him, consciousness, ignorant of its own origin, seizes upon itself in analytic self-penetration, suspended between subject and object, between meaning and being. The real world is not a thing founded in itself, that can in a significant manner be established as an independent existence. Recognition of the world as it comes from God cannot, as metaphysics and theology have repeatedly attempted, be achieved by cognitions crystallizing into separate judgments that have an independent meaning and assert definite facts. It can be gained only by symbolical construction. What this means will become clearer in the two following lectures.

Many people think that modern science is far removed from God. I find, on the contrary, that it is much more difficult today for the knowing person to approach God from history, from the spiritual side of the world, and from morals; for there we encounter the suffering and evil in the world which it is difficult to bring into harmony with an all-merciful and all-mighty God. In this domain we have evidently not yet succeeded in raising the veil with which our human nature covers the essence of things. But in our knowledge of physical nature we have penetrated so far that we can obtain a vision of the flawless harmony which is in conformity with sublime reason. Here is neither suffering nor evil nor deficiency, but perfection only. Nothing prevents us as scientists from taking part in the cosmic worship that found such powerful expression in the most glorious poem of the German language, the song of the archangels at the beginning of Goethe's *Faust*:

> The sun makes music as of old
> Amid the rival spheres of heaven
> On its predestined circle rolled
> With thunder speed; the angels even
> Draw strength from gazing at its glance,
> Though none its meaning fathom may:
> The world's unwithered countenance
> Is bright as on the earliest day.

> Die Sonne tönt nach alter Weise
> In Brudersphären Wettgesang,
> Und ihre vorgeschriebne Reise
> Vollendet sie mit Donnergang.
> Ihr Anblick gibt den Engeln Stärke,
> Wenn keiner sie ergründen mag;
> Die unbegreiflich hohen Werke
> Sind herrlich wie am ersten Tag.

II. Causality

Of the various ideas which, in the first lecture, were sketched rather than developed, we wish to consider causality in somewhat greater detail. This subject is also of vital interest in the natural science of the present, since modern quantum theory has precipitated a crisis of the concept of determination which dominated science in the last centuries.

Two relatively independent components appear to me to be fused in the idea of causality. I should like to designate them for the present quite generally as *the mathematical concept of determination by law*, and *the metaphysical notion of "the reason for something,"* that is the *Bestimmungsgrund*. Somewhere Leibniz says: "Just as the inner understanding of the word 'I' unlocks the concept of substance for me, so it is the observation of my self which yields other metaphysical concepts, such as 'cause,' 'effect,' and the like." The basic intuition through which we approach the essence of causality is: I do this. In this there is no question whatever of any kind of regularity—any kind of law—which holds again and again.

Descartes brings out the decisive point in the problem of free will with particular clarity, when he demonstrates the freedom involved in the theoretical acts of affirmation and negation: When I reason that $2 + 2 = 4$, this actual judgment is not forced upon me through blind natural causality (a view which would eliminate thinking as an act for which one can be held answerable) but something purely spiritual enters in: the circumstance that $2 + 2$ really equals 4, exercises a determining power over my judgment. The issue here is not that the determining factors responsible for my actions (in part) lie in me, as an existing natural being, and not outside of me; nor that entirely groundless, blind decisions are possible. But one has to acknowledge that the realm of Being, with respect to its determining factors, is not closed, but open toward mind in the ego, where meaning and being are merged in an indissoluble union. If I just now stated that the circumstance that $2 + 2 = 4$ exercises a power over my actual judgment, I did not thereby mean to imply a spiritual realm of facts or of Platonic ideas having an independent existence above reality, but I wished to emphasize that we are here dealing not with a new realm of existence but only with meaning—meaning which finds its fulfillment in reality.

The method of scientific research, primarily introduced by Galileo, presents two aspects, both equally essential, which are somewhat related to this juxtaposition of meaning and being: the a priori side, namely, free mathematical construction of the field of possibilities, and the a posteriori empirical side, the subjection of reality to experience and experiment. The history of the Renaissance shows very clearly how a positivistically inclined empiricism does not find in itself sufficient power to push through to a discovery of the natural law, but always sinks back into theosophy, mysticism, and magic. The approach of Leonardo and Galileo, who seek the reasons of reality in experience, is sharply separated from the ways of sensualistic doctrines; as the former clearly and definitely points toward mathematical idealism, so the latter always lead back to the primitive forms of animism; Campanella, also Cardano, and even Bacon are examples. On the other hand, through the great discoveries of Copernicus, Kepler, and Galileo, as well as the accompanying theories advocating the construction of nature through a priori given, logical-mathematical elements, there was established a supreme realization of the autonomy of the human intellect and its power over matter. In the philosophy of Descartes, which is the most universal expression of the thought of this epoch, the new mechanical interpretation must therefore be reconciled with the idealism of freedom; for an intensified consciousness of dignity and personal freedom resulted from that self-certainty of reason, which is so often and so naturally bound up with the constructive power of the mathematical mind. But for rational thinking, the duality of natural determination and personal freedom involved a serious antinomy, since the concrete person of the individual is embedded in nature.

It is well known that the first modern theory of determinism was carried. through by Hobbes. One of its clearest formulations we owe to Laplace. I quote his famous words from the *Essai philosophique sur les probabilités*:

> An intelligence which knows the forces acting in nature at a given instant, and the mutual positions of the natural bodies upon which they act, could, if it were furthermore sufficiently powerful to subject these data to mathematical analysis, condense into a single equation the motion of the largest heavenly bodies and of the lightest atoms; nothing would be uncertain for it, and the future as well as the past would lie open before its eyes. The human mind, in the perfection to which it has carried astronomy, offers a weak image of such an intelligence in a limited field.

If it is true that I am an existing individual performing real mental acts and at the same time a self-penetrating light, mind that is open toward meaning and reason, or, as Fichte expressed it, "force to which an eye has been lent"; and if Descartes' conviction of freedom is not deceptive—that is, if the realm of being with respect to its determining factors is not closed, but open toward reason in the ego—then this feature of openness must also manifest itself within nature and its science. Since this was not the case in natural science as it developed from Galileo's time with the native claim of embracing all of nature, this natural science became to the modern mind the power which shook the naive belief in the independence of the ego. Everything supports the fact that living beings do not violate the exact laws of nature; I, for example, can only impart a momentum to my body by pushing off from other bodies, which thereby take on an opposite momentum. Natural science is too easily condemned as rank materialism in view of its adherence, through many centuries, to a strictly deterministic position. Anyone aware of the extensive applicability and the precision of the mathematical laws of nature, as they were revealed principally by astronomy and physics, must admit that this position was the only fruitful one; the limits of determination by law will be discovered when one follows this way to its end, not, however, by giving way to evasive compromises, out of indolence or sentimentality. We firmly believe today that we have touched these limits in quantum mechanics.

After these preliminary remarks I now turn to the problem of the determination of nature by mathematically formulated laws. I shall begin with certain epistemological considerations concerning the meaning of the law of causality. Decisive as these considerations may be for the methodology of natural science, they accomplish little, I believe, in the way of relieving the pressure which a determination through the world of things places upon the ego. In the second part, however, we shall turn to the problem proper, in order to ascertain from concrete physics, as it developed in the last decades, the character of the determination it asserts and the limits of such determination.

The transformation of the metaphysical question of cause into the scientific question of law is taught by all great scientists. The discovery of the law of falling bodies is the first important example; Galileo himself says about it in his *Discorsi*: "It does not seem to me advantageous now to examine what the cause of acceleration is." It is more important to investigate the law according to which the acceleration varies. Again, Newton says:

> I have not yet been able to determine from the phenomena the cause of these properties of gravitation, and I do not invent hypotheses (*Hypotheses non fingo*). It is sufficient that gravitation exists, that it acts according to the laws we have formulated, and that it is capable of explaining all motions of heavenly bodies and of the sea. (End of *Principia*.)

Dynamics, according to the doctrines of d'Alembert and Lagrange, requires no laws which extend to the causes of physical phenomena and to the essence of such causes; it is closed in itself as a representation of the regularities of phenomena.

To be sure, the statement that the course of events is determined by means of natural laws does not exhaust the content of what appears to us, perhaps somewhat vaguely, as the relation of cause and effect. In particular, the mathematical law cannot distinguish between the determining and the determined. If several quantities a, b, c are functionally related, for example, $a + b = c$, then the value of a and b may determine that of c; but the same law may also be so construed that, by means of the quantities a and c, it determines b. If natural laws enable us to predetermine the future, we can, with their help, equally well determine the past from the present. The general law of refraction of light in an optically inhomogeneous medium, as, for example, the atmosphere, may, according to Snellius [Snell], be formulated as a differential law which connects the infinitesimal change in the direction of a light ray with the change in the velocity of propagation along the ray. But as an alternative we may, according to Fermat, describe the same process by reference to the principle that, in passing from one point of the medium to a distant point, the ray chooses that path which requires the least time. The differential formulation corresponds to the causal conception according to which the state at one instant determines the change of state during an infinitesimal interval of time; the second, the integral formulation, savors of teleology. However, both laws are mathematically equivalent. Thus natural law is completely indifferent to causality and finality; this difference does not concern scientific knowledge, but metaphysical interpretation by means of the idea of determining reason. I believe it is necessary to state this with full clarity: the law of nature offers as little evidence for or against a metaphysical-teleological interpretation of the world as it does for or against a metaphysical-causal one.

The first epistemological analysis of the law of causality aiming to isolate that part of causality which plays a role in an actual investigation of nature was undertaken by Hume. As preliminary characteristics he finds: (1) The principle of nearby action, according to which causally related objects or processes must be directly connected in space-time; the answer to the question "Why?" demands the insertion of a continuous uninterrupted causal chain. (2) The transition: cause $\rightarrow$ effect runs in the time sense: past $\rightarrow$ future. (3) The necessity of the causal bond which is commonly postulated, and which is taken over from the idea of fate, is, according to Hume, not capable of a clear-cut empirical interpretation. He therefore replaces necessity by repetition and permanence; that is, whenever the same circumstances recur, the same effect will follow the same cause. But even with this nothing is gained, as an event happens in its full concretion only once. It is thus necessary that certain demands of continuity be added, stipulating that causes differing sufficiently little from one another have effects also differing but little; that sufficiently remote bodies or events have a negligible effect, and so on. The phenomena must be brought under the heading of concepts; they must be united into classes determined by typical characteristics. Thus the causal judgment, "When I put my hand in the fire I burn myself," concerns a typical performance described by the words "to put one's hand in the fire," not an individual act in which the motion of the hand and that of the flames is determined in the minutest detail. The causal relation therefore does not exist between events but between types of events. First of all—and this point does not seem to have been sufficiently emphasized by Hume—*generally valid relations must be isolated by decomposing the one existing world into simple, always recurring elements.* The formula "*dissecare naturam* [to dissect nature]" was already set up by Bacon.[9]

I do not intend to go into the details of an analysis of nature, but shall direct attention to only two or three points. (1) One does not hesitate to decompose hypothetically things that are irreducible simple elements from a perceptual standpoint, as, for example, the white sunlight into the spectral colors, or the acceleration which the earth acquires into the partial acceleration which the sun and the planets separately impart to it. (2) In scientific investigation one does not stop with the perceived qualities of a body which directly appeal to the senses, but one introduces "concealed characters" which only manifest themselves through the reactions of that body with others. Thus, for example, the inertial mass is no perceivable characteristic of a body, but can only be determined by allowing the body to react with others and then applying the impulse law to these reactions. This law asserts: to every isolated body a momentum may be assigned, this momentum being a vector with the same direction as the velocity; the positive factor m, by which the velocity must be multiplied in order to give the momentum, is called the mass. If several bodies react on each other, the sum of their momenta after the reaction is the same as before. It is only through this law that the concept of momentum, and with it that of mass, attains a definite content; separated from it they are simply suspended in the air. It is this constructive method alone which permitted natural science to penetrate beyond the narrow bounds of the purely geometrical concepts, within which Descartes attempted to confine it. Even the geometrical concepts have essentially this constructive character. (3) It is typical of the mathematizing sciences (in contradistinction to the descriptive ones) that they pass from the classification of given examples, like Linnaeus' classification of the actually occurring plants, to the ideal, constructive generation of the possible. Instead of classifying the perceivable colors, physics sets up the concept of ether waves, which may differ only in direction and wave length. Both direction and wave length, however, vary within a predetermined domain of possibilities. Thus the four-dimensional medium of space and time is the field of possible coincidences of events. Such a field, and a most important one, open to our free construction, is the continuum of numbers. To be sure, the analysis must be carried to the point where each element may be determined, in its full concretion, through particular values of such constructive moments as direction and wave length, which vary within a domain completely surveyable since it arises from free construction. The law of causality then maintains that between such quantitative elements there exist universally valid, simple, exact, functional relations.

Let us now pass from elementary analysis to the idea of natural law. Is it so self-evident that it requires no further exposition? I think not. Above all I wish here to emphasize two points.

The assertion that nature is governed by strict laws is devoid of all content if we do not add the statement that it is governed by mathematically simple laws. This matter is somewhat analogous to the fundamental law of multiple proportions in chemistry: it loses all its content unless we add that the combination occurs in small integral multiples of the relative atomic weights. That the notion of law becomes empty when an arbitrary complication is permitted was already pointed out by Leibniz in his *Metaphysical Treatise*. Thus simplicity becomes a working principle in the natural sciences. If a set of observations giving the dependence of a quantity y on a quantity x lie on a straight line when plotted, we anticipate, on account of the mathematical simplicity of the straight line, that it will represent the exact law of dependence; we are then able to extrapolate and make predictions. One cannot help but admit that this working principle of simplicity has stood the test well. Euclidean geometry, for example, as a science concerning the metric behavior of rigid bodies, was gained from very rough experiences as their simplest interpretation. In later precise geometrical and astronomical measurements this geometry proved to hold much more exactly than we could have anticipated from its origin. Analogous cases are continually encountered in physics. The astonishing thing is not that there exist natural laws, but that the further the analysis proceeds, the finer the details, the finer the elements to which the phenomena are reduced,

the simpler—and not the more complicated, as one would originally expect—
the fundamental relations become and the more exactly do they describe the
actual occurrences. But this circumstance is apt to weaken the metaphysical
power of determinism, since it makes the meaning of natural law depend on
the fluctuating distinction between mathematically simple and complicated
functions or classes of functions.

In the same direction points the epistemological observation that the principle, "under the same circumstances the same results will follow" (no matter
how one may interpret it), does not hold as something verifiable by experience. An inductive proof of the proposition, as Helmholtz says, would be very
shaky; the degree of validity would at best be comparable with that of the
meteorological rules. It is rather a norm whose validity we enforce in building up our experience. This is well illustrated by the example of the spectral
analysis of white light by means of a prism, to which we referred previously.
In obvious contradiction to the fundamental proposition that under equal circumstances equal causes will call forth equal reactions, two colors which appear
as the same white to the senses yield totally different spectra, in general, after
passing through the same prism. In order to save our fundamental proposition
we invent a "hidden" variety in white light, which is most suitably described by
giving the spectrum itself with its intensity of distribution; it is for this reason
that we are led, in physics, to regard simple white light as a composite of colors.
(We note that here at first the apparatus used in the reaction, the prism with its
special properties, still plays a role; it is only after varying the shape, substance,
and orientation of the prism with respect to the light rays, and thus separating
the two influences from one another, that one arrives at a scale of wave lengths
which is independent of the prism.)

Constructive natural science is confronted with the general problem of
assigning to objects such constructive characteristics that their behavior under
circumstances described by the same kind of characteristics is entirely determined and predictable by means of the natural laws. The implicit definition
of the characteristics is bound to these laws. The fact that we do not find but
enforce the general principles of natural knowledge was particularly emphasized
by the conventionalism of H. Poincaré.[10] But I believe one may also consider
the hastily sketched developments just completed as an interpretation of Kant's
doctrine of the categories.

These considerations force upon us the impression that the law of causality as a principle of natural science is one incapable of formulation in a few

words, and is not a self-contained exact law. Its content can in fact only be made clear in connection with a complete phenomenological description of how reality constitutes itself from the immediate data of consciousness. Kant's naive formulation: "Everything that happens (comes into existence) implies something from which it follows according to a rule," can hardly satisfy us any longer. At the same time "fate" as expressed in the natural laws appears to be so weakened by our analysis that only through misunderstanding can it be placed in opposition to free will.

True as this may be with respect to the general principle of causality, as a methodical principle of natural science, yet I believe that this epistemological subterfuge, so eagerly adopted by just the deeper thinkers, is invalidated by concrete physics itself. So far we have spoken only of the methodology of natural science and its leading principles. But through it results concrete physics itself, deeply rooted in the fertile soil of experience. Perhaps there is no strict logical way leading from the facts to our theories; but physics as a whole is convincing for everyone who devotes himself seriously to an investigation of the cosmos. It is now no longer a question of the general idea of the mathematically simple natural law, but the definite concrete laws of nature themselves stand before us in their wonderfully transparent mathematical harmony. The previous decomposition of the world into individual systems, individual events and their elements vanishes more and more as the theoretical structure is completed; the world appears again as a whole, with all its parts interactively bound to one another. The development tends distinctly toward a unified, all-embracing world law. In the actually known natural law lies a restriction of the world structure which in all metaphysical seriousness sets a limit even to the claims of autonomy of the mind. Therefore we shall now concern ourselves with this lawfulness itself, to see how it is constituted and where its limits are.

A first consideration is this: physics has never given support to that truly consistent determinism which maintains the unconditioned necessity of everything which happens. Even from its most extreme standpoints, including Newton's physics of central forces as well as modern field-theory, physics always supposed the state of the world at a certain moment in a section $t = $ const. [constant] to be arbitrary and unrestricted by laws. Even in Laplace's universe there was an "open place" which could be chosen at random among the sections $t = $ const. of the world. This perhaps suffices to reconcile mechanical necessity with Divine Predestination. Descartes argues thus: since neither the nature nor the distribution of the material constituents of the world nor their initial velocities are to be derived by pure reasoning, God could have set up the natural order in innumerable ways; He chose one to suit His purpose. Newton makes similar remarks in the conclusion of his *Opticks*. But this degree of arbitrariness seems to me insufficient to admit human free will. My own destiny in the world from birth to death could still, on this view, be fixed by the state of the world in a time-section which has no contact with my existence, with the world-line of my life, since it precedes or follows it. Hence Kant's solution of the dilemma (the meaning of which was so vague even to himself that he found difficulties in understanding the changes of human character) can only be carried through honestly if one believes in the existence of the individual from eternity to eternity, in the form of a Leibniz monad, say, or by metempsychosis as the Indians and Schopenhauer believe. Nevertheless, it is of sufficient importance that physics has always admitted a loophole in the necessity of Nature.

The antinomy between freedom and determination takes its most acute form in the relation between knowing and being. Let us assume once more with Laplace that the state of the world at one moment, i.e., a three-dimensional section $t = $ const., defines by strict mathematical laws its course during all past and future time. Then we might suppose that I can calculate the future from what I know (or can know) here and now at the world point O. I should like to state with all emphasis that this antinomy, which formerly existed, disappears in the relativity theory. In the first lecture I described the causal structure according to which a kind of conical surface issues from each point O of the four-dimensional world as vertex and separates the causal past and future. Causality is here not merely a methodological principle but becomes through this structure an objective constituent of the world. In the figure the section $t = $ const. through O separates the past and the future sheets of the cone through O. But it is not this plane section, it is the surface of the backward light-cone which separates what is knowable at O from what is not. And it is a mathematical consequence of the classical physical laws that whereas the backward half of the world, cut off by $t = $ const., determines the whole, the interior of the backward light-cone does not. That is to say, only after a deed is done can I know all its causal premises.

If we regard, however, our problem as concerning reality alone and not concerning the relation of knowledge and reality, and if free action shall be possible in this real world, then we must demand that the content of the forward pointing cone through O shall not be completely determined by the rest of the world. This would contradict classical physics. But classical physics, after decades of invasion by statistical theories, is now finally superseded by the quantum theory, and a new situation has arisen.

In three grams of hydrogen there are about 10^{24} hydrogen molecules whirling about; it is of course impossible to calculate exactly their motion under the forces they experience from the walls of the container and from one another. Their average velocity determines the temperature, their bombardment of the walls, or rather the impulse per unit area it conveys, the pressure. Certain mean values are what our observations measure and these can be predicted by probability calculations, without detailed investigation of the motion. Consider, for example, a cubical container divided up into many small cubes, all of equal size, and suppose the chance of a given molecule to be in one of these is the same for each and that the space probabilities of the different molecules are independent in the statistical sense. Then we can show that the gas density in each of the small cubes differs with utterly overwhelming probability by less than, say, 0.01% from the mean density of the whole. Macroscopically speaking, the gas in equilibrium is uniformly dense. In the same way the kinetic theory of gases, first formulated by Daniel Bernoulli, leads to the other well-known gas laws.

The theory of probability not only tells us the mean value of a quantity, but also how great its deviation from this mean may be expected to be. The spontaneous variations in the density of the atmosphere which arise through the random motions of its molecules are the cause of the diffusion of the sun's rays in daylight, which makes a cloudless sky appear not black but blue. Small though they are individually, combined they produce a perceptible effect. Such variation-phenomena are the main supports of the statistical theory.[11] The powerful researches of Maxwell and Boltzmann have made clear that the majority of physical concepts are not exact in the sense of classical physics, but statistical mean values, with a certain degree of indetermination, and that most of the familiar laws of physics, especially all those which concern the thermodynamics of atomic matter, are not to be regarded as strictly valid natural laws but as statistical regularities.

The first epistemological attitude toward statistical physics was to regard the probability theory simply as a short cut to certain consequences of the exact laws. For instance, strictly speaking, one would have to prove by means of the classical laws of motion that the time intervals during which the gas deviates noticeably from thermodynamic equilibrium were together vanishingly small as compared to the whole period of observation. Attempts at such proofs were indeed made, but it was always necessary to introduce an unproved hypothesis, the so-called ergodic hypothesis, at the critical point.[12] If we adhere to the actual practice of physical research we are bound to admit that with the progress of the statistical theory and its continual increase in fruitfulness the attempts to base it on strict functional laws have gradually been abandoned. Historical evolution has spoken and demands that we recognize statistical concepts as equally fundamental with the concepts of law. I believe that such historical evolution can exert a more compelling pressure than any reasoning which pretends to be heaven knows how rigorous.

It should be remarked in this connection that in the world of exact laws time is reversible; changing t into $-t$ makes no difference. On the other hand, the definite direction of flow from past to future is perhaps the one outstanding mark of subjective time. This uniqueness of direction enters into physics not through its functional laws, but through our probability judgments; from a state at a given moment we deduce the probable state at a subsequent moment according to computed probabilities, and not the state at a previous one. Thus probability exposes a part of the causal idea which was quite suppressed in the exact laws.

Yet only the latest aspect of physics, quantum mechanics, has reduced the statistical nature of physical lawfulness to its ultimate foundations. This step became necessary in order to give an account of the double nature of physical entities, brought into evidence first in the case of light. Light is a spatially continuous undulatory process of electromagnetic nature. Only this conception enables us to understand diffraction and interference. But on the other hand a number of phenomena discovered in the last decades force us to conceive of light as consisting of single quanta, thrown out from the source of light in definite directions, and whose energy content is determined by the frequency, or the color of the light. I will describe here one of these phenomena. If a plate of metal is irradiated with ultra-violet light, electrons are emitted from the plate. Assuming the intensity of the light to be small, the energy of the wave which traverses an atom would not suffice to remove an electron from the atomic system. Even if we imagine some kind of a mechanism allowing the accumulation of wave energy within the atom, this effect could only begin after a long period of accumulation. Instead of this, it sets in immediately. The force with which the electrons are knocked out is totally independent of the intensity of the light; but it depends on its color. Only the number of electrons emitted in unit time increases with the intensity. This process can only be understood if light consists of single quanta. The energy content of such a light quantum, which hits an atom, is carried over to an electron, thus enabling this electron to break its bond with the nucleus of the atom, and furthermore imparting to the electron a certain kinetic energy. This energy depends on the energy content of the light quantum and hence on the color of the light. The dual nature of light—its being a wave capable of interference and also at the same time a light quantum striking suddenly here and there— we try to cover by assuming that the intensity of the wave field at a certain point represents the relative probability that a light quantum will be at that point. The more intense the light, the denser the accumulation of light quanta in unit time. The wave field obeys a strict functional law.[13] But exactly the same condition prevails for the constituents of matter, the electrons. Everyday experiences suggest that their nature is corpuscular. But electrons have lately been shown to be susceptible of diffraction and interference. Hence there exist precise laws, but they deal with wave fields and therefore with quantities, which for real events have only the significance of probabilities. They determine the actual processes in the same way that a priori probabilities determine statistical mean values, frequencies—always containing a factor of uncertainty.

You know how it is possible with the aid of a prism or a grating to select monochromatic light from natural light. All light quanta in a ray of monochromatic light have the same definite energy and the same momentum. If we let the ray traverse a Nicol prism, we impress on it a certain direction of polarization.[14] Let us describe this in terms of light quanta. A certain light quantum either will pass through the Nicol or it will not; hence there may be ascribed to the light quantum a certain quantity q_s, corresponding to the position s of the Nicol, and taking on the values $+1$ or -1, according as the light quantum passes through or not. The monochromatic, polarized, plane light wave is the utmost in homogeneity that is obtainable. But we observe that such a homogeneous ray of light is again split up into a transmitted and reflected ray, when sent through a second Nicol in a position t different from s. The relative intensities are completely determined by the angle between the two positions s and t. They represent the probabilities that for a light quantum with $q_s = 1$, the quantity $q_t = +1$ or -1. The ray of light which passed through both Nicols is not more homogeneous than the ray which passed through only the first one: it is of exactly the same character as it would have been if we had omitted the first Nicol. Hence the selection due to the first Nicol is destroyed by the second one. It is legitimate to speak of the quantity q_s for a light quantum, because there exists a method of determining its value. We can also speak of the quantity q_t. But it is meaningless to ask for the values taken simultaneously by the quantities q_s, q_t for a light quantum, because measuring q_t by selecting the light quanta with $q_t = 1$ destroys the possibility of measuring q_s by selecting the light quanta with $q_s = 1$.

This impossibility is not due to human limitation, but must be regarded as an essential one. Another example will make this clearer.[15] An atom of silver possesses a certain magnetic moment, it is a small magnet of definite strength and direction. It can be represented by an arrow, the vector of magnetic moment. This vector has, in any spatial direction z, a component m_z capable of taking on only two values, ± 1, when measured in a certain unit, the magneton. By means of a magnetic field inhomogeneous in the direction z, it is possible to separate from a beam of atoms flying through the field the two component beams for which m_z equals $+1$ and -1 respectively. The same evidently applies in any other spatial direction. But a vector, whose components in every spatial direction are capable of taking on only the values ± 1, is geometrically absurd. The resolution of this paradox is this: if the component m_z is fixed by the separation, then no further component can be determined. Only probabilities can be calculated for their possible values ± 1.

Classical physics in attempting to establish conditions which would guarantee maximum homogeneity, assumed that for such a "pure case" any physical quantity of the physical system considered took on a well defined value, which under the same conditions would always be reproduced. Quantum mechanics also requires the experimenter to create a pure case whose homogeneity cannot be increased. But the ideal of classical physics is not realizable for quantum mechanics. We must not ask what value is taken on by a physical quantity in a certain pure case, but instead what the probability is that this physical quantity will take on a given value in this pure case. The idea that an electron describes a path cannot be upheld any longer. It is true that an electron's position at a certain instant can be measured; its velocity, too, is measurable, but not both at the same time. The measurement of position destroys the possibility of an exact measurement of speed. There is no human incapacity involved; the difficulty lies in the very nature of things. The meaning of a physical quantity is bound to the method by which it is measured. The attributes with which physics deals manifest themselves only through experiments and reactions which are based on postulated laws of nature. Formerly physicists took the point of view that these attributes were assigned to the physical bodies themselves, independently of whether or not the measurements necessary to establish them were actually carried out. It was proper to connect them by the logical "and"; it was reasonable to postulate determinism and to satisfy this methodical postulate by introducing suitably chosen, concealed attributes. This epistemological position of constructive science is now submitted to an essential restriction in quantum mechanics.

We may try to escape this verdict by saying that the wave field, which obeys precise laws, is reality. Nevertheless it is a fact that this wave field cannot be observed directly, but only determines all observable quantities in the same way that a priori probabilities determine statistical frequencies. In this connection the uncertainty principle is unavoidable. We may say that there exists a world, causally closed and controlled by precise laws, but in order that I, the observing person, may come in contact with its actual existence, it must open itself to me. The connection between that abstract world beyond and the one which I directly perceive is necessarily of a statistical nature. This fact, together with the new insight which modern physics affords into the relation between subject and object, opens several ways of reconciling personal freedom with natural law. It would be premature, however, to propose a definite and complete solution of the problem. One of the great differences between the scientist and the impatient philosopher is that the scientist bides his time. We must await the further development of science, perhaps for centuries, perhaps for thousands of years, before we can design a true and detailed picture of the interwoven texture of Matter, Life and Soul. But the old classical determinism of Hobbes and Laplace need not oppress us any longer.

Another feature of quantum mechanics is worth mentioning. The state of a physical system is determined when for each physical quantity of the system the probability of its taking on each possible value is known. It is true therefore that the state of a system consisting of two electrons determines the states of both electrons, but the converse does not follow. The knowledge of the states of the two parts of a system by no means fixes the state of the whole system. We find here a definite and far-reaching verification of the principle that the whole is more than the sum of its parts. Modern vitalism, among whose proponents I mention first of all Driesch, has attempted to reduce the independence of life, its essential distinction from non-organic processes, to the concepts of Gestalt or the Whole. According to vitalism the living organism reacts as a whole;

its functions are not additive. The manner in which its structure is preserved throughout growth, in spite of all outside influences and perturbations, is not to be explained by small scale causal reactions between the elementary parts of the organism. Now we see that according to quantum physics the same applies even to inorganic nature and is not peculiar to organic processes. It is out of the question to derive the state of the whole from the state of its parts. This leads to conditions which may most plainly if not most correctly be interpreted as a peculiar non-causal "understanding" between the elementary particles, that is prior to and independent of the control exercised by differential laws which regulate probabilities. The rule of W. Pauli that two electrons may never be found in the same quantum state is one of the best illustrations. It seems therefore that the quantum theory is called upon to bridge the gap between inorganic and organic nature; to join them in the sense of placing the origin of those phenomena which confront us in the fully developed organism as Life, Soul and Will back in the same original order of nature to which atoms and electrons also are subject. So today less than ever do we need to doubt the objective unity of the whole of nature, less than ever to despair of attaining unity of method in all natural sciences.

III. Infinity

In the first lecture I pointed out that the Greeks made the divergence between the finite and the infinite fruitful for the understanding of reality, and that this is one of their greatest achievements. To illustrate how the early Greek thinkers formulated the notion of the infinite in a manner enabling it to bear upon science, I shall start with a fragment transmitted to us from Anaxagoras: "In the small there is no smallest, but there is always still a smaller. For what is can never cease to exist through division, no matter how far this process be pursued." This statement, of course, refers to space or a body. The continuum, Anaxagoras says, cannot be composed of discrete elements which are "chopped off from one another, as it were, with a hatchet." Space is not only infinite in the sense that in it one nowhere reaches an end; but at every place it is infinite if one proceeds inward toward the small. A point can only be identified more and more precisely by the successive stages of a process of division continued ad infinitum. This is in contrast with the state of immobile and completed being in which space appears to direct perception. For the *quale* [quality] filling it, space is the principle of distinction which primally creates the possibility for a diverseness of qualitative character; but space is at the same time distinction and contact, continuous connection, so that no piece can be "chopped off . . . with a hatchet." Hence a real spatial thing can never be given adequately; it unfolds its "inner horizon" in an infinite process of continually new and more precise experiences. Consequently it appears impossible to postulate a real thing as being, as closed and complete in itself. In this manner the problem of the continuum becomes the motive for an epistemological idealism: Leibniz, among others, testifies that it was the search for a way out of the "labyrinth of the continuum" which first led him to conceive of space and time as orders of the phenomena. "From the fact that a body cannot be decomposed mathematically into primal elements," he says, "it follows immediately that it is nothing substantial but only an ideal construction designating merely a possibility of parts, but by no means anything real."

Anaxagoras is opposed by the strictly atomistic theory of Democritus. One of his arguments against the unlimited divisibility of bodies is approximately as follows: "It is contended that division is possible; very well, let it be performed. What remains? No bodies; for these could be divided still further, and the division would not have progressed to the ultimate stage. There could only be points, and the body would have to be composed of points, which is evidently absurd." The impossibility of conceiving the continuum as in a stage of rigid being cannot be illustrated more pregnantly than by Zeno's familiar paradox about the race of Achilles with the tortoise. The tortoise has a start of length 1; if Achilles moves with twice the speed of the tortoise, the tortoise will be the distance $\frac{1}{2}$ ahead of him at the moment when Achilles arrives at its starting point. When Achilles has covered this distance also, the tortoise will have completed a path of length $\frac{1}{4}$, and so on, ad infinitum; whence it is to be concluded that the swift-footed Achilles never catches up with the reptile.

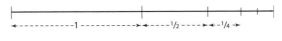

Figure 4.3

The observation that the successive partial sums of the series

$$1 + \frac{1}{2} + \frac{1}{4} + \cdots$$

do not grow beyond all bounds but converge toward 2, by means of which the paradox is thought to be done away with today, is certainly important, pertinent and elucidating. But if the distance of length 2 really consists of infinitely many partial distances of length $1, \frac{1}{2}, \frac{1}{4}, \ldots$ as "chopped off" integral parts, then it contradicts the essence of the infinite, the "incompletable," to say that Achilles has finally run through them all. Aristotle remarks with reference to the solution of Zeno's paradox, that "the moving does not move by counting," or more precisely:

> If the continuous line is divided into two halves, the one dividing point is taken for two; it is both beginning and end. But as one divides in this manner, neither the line nor the motion are any longer continuous. . . . In the continuous there is indeed an unlimited number of halves, but only in possibility, not in reality.

Since Leibniz seeks the foundation of the phenomena in a world of absolute substances, he has to accept the stringent argumentation of Democritus; he conceives the idea of the monad. He says, in agreement with Aristotle:

> In the ideal or the continuum the whole precedes the parts. . . . Here the parts are only potential. But in substantial things the simple precedes the aggregates, and the parts are given actually and before the whole. These considerations resolve the difficulties concerning the continuum, difficulties that arise only if one considers the continuum as something real which in itself has real parts prior to any division performed by us, and if one regards matter as a substance.

This suggests, as a solution of the antinomy of the continuum, the distinction between actuality and potentiality, between being and possibility. The application of mathematical construction to reality then ultimately rests on the double nature of reality, its subjective and objective aspect: that reality is not a thing in itself, but a thing appearing to a mental ego. If we assume Plato's metaphysical doctrine and let the image appearing to consciousness result from the concurrence of a "motion" issuing partly from the ego and partly from the object, then extension, the perceptual form of space and time as the qualitatively undifferentiated field of free possibilities, must be placed on the side of the ego. Mathematics is not the rigid and uninspiring schematism which the layman is so apt to see in it; on the contrary, we stand in mathematics precisely at that point of intersection of limitation and freedom which is the essence of man himself.

If now we proceed to formulate these old ideas a little more precisely, we first discover the infinite in a form more primitive than that of the continuum, namely in the sequence of natural numbers $1, 2, 3, \ldots$; and only with their help can we begin to attack the problem of the mathematical description of the continuum. Four stages can be distinguished in the development of arithmetic as regards the part played by the infinite. The first stage is characterized by individual concrete judgments, like $2 < 3$, the number symbol $//$ is contained in the symbol $///$. In the second stage there appears, for example, the idea of $<$, of "being contained" for arbitrary number symbols; and also the proposition of hypothetic generality: if any two number symbols a, b, are given, either $a = b$, or $a < b$, or $b < a$. The domain of the actually given is hereby not transgressed, since the assertion purports to be valid only when definite numbers are given. Something entirely new, however, takes place in the third stage, when I embed the actually occurring number symbols in the sequence of all possible numbers, which originates by means of a generating process in accordance with the principle that from a given number n a new one, the following one n', can always be generated. Here being is projected onto the background of the possible, or more precisely into an ordered manifold of possibilities producible according to a fixed procedure and open into infinity. Methodically this standpoint finds its expression in the definition and conclusion by complete induction. The principle of complete induction states that in order to establish that a property P relating to an arbitrary natural number n belongs to every such number, it is sufficient to prove: (α) 1 has the property P; (β) if n is any number having the property P, the following number n' also has the property P. The familiar method of distinguishing the even and the odd numbers from one another by "counting off two at a time" is a simple example of the definition by complete induction; it can be put in the form: (α) 1 is odd; (β) according as n is even or odd, n' is odd or even.

At this stage, the general statements of the science of numbers deal with the freedom of bringing the sequence of numbers to a stop at an arbitrary place. This consummates the transition to theoretical cognition proper: the transition from the a posteriori description of the actually given to the a priori construction of the possible. The given is embedded in the ordered manifold of the possible, not on the basis of descriptive characteristics, but on the basis of certain mental or physical operations and reactions to be performed on it—as, for example, the process of counting. The fourth stage of arithmetic will not be discussed in detail until later. It is the stage in which, following the prototype of the Platonic doctrine of ideas, the possible is converted into transcendental and absolute being, in its totality naturally inaccessible to our intuition.

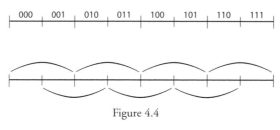

Figure 4.4

For the moment we refrain from taking this dangerous step and turn from the natural numbers to the continuum, asking how it must be described as the substratum of possible divisions which are continued ad infinitum. As an example I take the one-dimensional segment. I divide it into two pieces by a point (division of the first stage); in the second stage of division each of them is again decomposed into two pieces by a dividing point, so that we now have four pieces. In the division of the third stage, each of them is again divided into two, and so on, ad infinitum. At every step of the division, the number of pieces rises to twice its previous value; after the nth stage of division it amounts to 2^n. This is the method of diaeresis [division] by means of which Plato attempted to build up his ideal numbers. In the succession of divisions of the first, second, third, . . . , nth, . . . stage, we encounter the developing infinite sequence of numbers. I should like to prescribe that the division shall always be a bisection. But as long as I adhere to the intuitive nature of the continuum, I am prohibited from doing so; while by its very nature the continuum is divisible, the limits of the division can never be set exactly, although there is the possibility of improving the exactness and fineness of the division by continuing it to higher stages indefinitely. Hence we can at first set the limits only with a certain vagueness, but we must imagine that as we progress to the more advanced stages of the division, the division points of the preceding stages are fixed more and more precisely. Here we have to do with a process of "becoming" which in a really given continuum can only be performed up to a certain stage. But from the performance of this process on a concrete continuum, we can abstract its arithmetic scheme, and this is determined into infinity; this scheme is the subject matter of the mathematical theory of the continuum. In order to describe the arithmetic structure of the division, one must characterize the successively formed pieces in a systematic manner by symbols and indicate by means of these symbols how the pieces of the nth division stage adjoin, and how these pieces are formed from those of the preceding stage by the nth division. If the left half is always characterized by 0 and the right half by 1, the following divisional scheme is obtained: One can also consider the simple

sequence of natural numbers as such a scheme of division: here an undivided entity is decomposed in one piece (the 1) which is retained as a unit, and an undivided remainder; the remain-

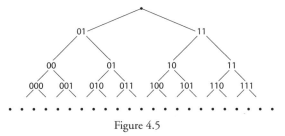

Figure 4.5

der is again decomposed in one piece (2) and an undivided remainder, and so on. The most illustrative realization of this process is time, as it is open into the future and again and again a fragment of it is lived through. Here not every part but only the last remainder is always subjected to bipartition. This is a simpler divisional scheme than that of the continuum, yet in principle it is of the same kind.

We combine every two adjacent parts of the nth stage to form a dual interval of the nth stage. (See fig. 4.4.) These intervals overlap in such a manner that if a point is known with sufficient accuracy, one can with certainty indicate one of the dual intervals of the nth stage in which it falls. An individual place is thus fixed more and more precisely, caught by an infinite sequence of dual intervals each of which lies entirely within the preceding one. This process is in principle equivalent to the one taught by Eudoxos in antiquity and used to locate points in the continuum and to distinguish them from one another. What modern times have added is the recognition of the fact that the sequence must not be considered only as a means for describing the location of a given point the existence of which is secured independently, but that it originally generates the point in the continuum constructively. Every such sequence furnishes a point, and in the arithmetic scheme the points are created by this procedure. It is only on the basis of this constructive turn that a mathematical mastery, an analysis, of continuity is possible.

But there is still a problem in the idea of the infinite sequence. All essential features of our analysis are preserved and the conditions are a little easier to describe if we take as its basis the sequence of natural numbers instead of the scheme of continued bisection: A sequence of natural numbers is being formed if I arbitrarily choose a first number, then a second, a third, and so forth: "free sequence of choice." But statements concerning this sequence have meaning only if their truth can be decided at a finite stage of the development. For example, we may ask if the number 1 occurs among the numbers of the sequence up to the 10th stage, but not whether 1 occurs at all, since the sequence never reaches completion. An individual definite sequence, determined into infinity, cannot be produced in this way through definite choice, but a law is necessary which allows us to calculate in general from the arbitrary natural number n the number occurring at the nth place of the sequence. If one adheres to this scheme, one sees that, in agreement with a remark of Aristotle quoted above, it is quite impossible to decompose the continuous segment $0 \ldots 1$ into two parts $0 \ldots \frac{1}{2}, \frac{1}{2} \ldots 1$ in such a manner that every point x belongs to either the one or the other of the two halves.

We have now dealt with the infinite in two forms: (1) the free possibility of bringing the sequence of numbers $1, 2, 3, \ldots$ to a stop at an arbitrary place (single act of choice); (2) the free possibility of forming a continually developing and never ending sequence of natural numbers (act of choice repeated ad infinitum), which, however, turns into a law when it is to represent a special sequence determined into infinity.

Before the discovery of the irrational by Pythagoras or the mathematicians of the Pythagorean school, as long as only fractions had been used in measuring segments, the opinion prevailed that an individual point in the continuum could be fixed by one or two natural numbers, numerator and denominator of a fraction; the infinite (2) was thereby reduced to (1). But that would mean that an arbitrary sequence of numbers $\mathcal{L}$ would by a lawful rule determine a natural number $n_{\mathcal{L}}$, which characterizes the sequence itself in a unique way like a name. This is evidently impossible. The name $n_{\mathcal{L}}$ must be determined when the development of the sequence $\mathcal{L}$ has reached a certain place. It would not have to be a fixed place that can be indicated in advance, for example the 2nd or the 100,000th, but could depend on the outcome of the acts of choice; eventually, however, $n_{\mathcal{L}}$ must be fixed and then cannot be altered by the further development of the sequence. But then all sequences having the same beginning up to this place furnish the same $n_{\mathcal{L}}$ no matter how they may differ from one another in their further development.

In this exposition I have so far largely followed the Dutch mathematician Brouwer who has in our day rigorously followed out the intuitional standpoint in mathematics.[16] This standpoint emphasizes the conflict between being and possibility. Metaphysics has at all times tried to overcome the dualism between subject and object, being and possibility, existence and meaning, limitation and freedom. At the end of the first lecture, I expressed my conviction that the origin and the reconciliation of this divergence can lie only in God. The attempt presented by realism to elevate the object to the dignity of absolute being was doomed to fail from the start; as was also the opposite attempt of idealism to endow the subject with the same high independence. In mathematics also, partly through its dependence on philosophy, the inclination toward the absolute which is evidently deeply rooted in man has asserted itself. Having described the infinite in mathematics under the category of possibility in the first part of my lecture, I want now to discuss the attempts to convert the field of possibilities that is open into infinity into a closed domain of absolute existence. Four different attempts to reach this goal stand out in the course of history. The two older ones really have to do with the continuum only.

The first and most radical attempt makes the continuum consist of countable discrete elements, atoms. This procedure was followed already in antiquity by Democritus in explaining the nature of matter, and has met with most brilliant success in modern physics. Plato, clearly conscious of his proposed goal—the salvation of the phenomenon through the idea, seems to have been the first to conceive a consistent atomism of space. The atomistic theory of space was renewed in Islamic philosophy by the Mutakallimun, in the Occident by Giordano Bruno's doctrine of the minimum.[17] Revived by the quantum theory this idea reappears in our time in discussions about the foundations of physics. But so far it has always remained pure speculation and has never found the least contact with reality.

The second attempt deals with the infinitely small. The tangent to a curve at the point P is considered as the line joining P to the infinitely neighboring point of the curve, not as the limiting position which the secant PQ approaches indefinitely as the point Q converges toward P along the curve; velocity is the quotient of the infinitely small segment described in the infinitely small time dt, divided by dt, not the limit to which the corresponding quotient formed for a finite time interval converges when one lets the length of the time interval decrease below every bound.[18] Galileo compares the bending of a line into a regular polygon of a thousand sides with winding it on a circle; he considers this actually equivalent to the bending of the line into a polygon of infinitely many, infinitely small sides, although the individual sides cannot be separated from one another.[19] Condensation and rarefaction of matter are interpreted by him as an intermixing of infinitely small filled and empty parts of space in changing proportions. Although Eudoxos had rejected the infinitely small by means of a rigorously formulated axiom, this idea, vague and incomprehensible, becomes in the eighteenth century the foundation of infinitesimal calculus. The founders themselves, Newton and Leibniz, expressed with some degree of clarity the correct conception that this calculus does not deal with a fixed infinitely small quantity but a transition toward the limit zero; but this conception is not the guiding principle in the subsequent development of their ideas, and they evidently ignore the fact that the limiting process not only has to determine the value of the limit, but must first guarantee its existence. For this reason modern infinitesimal calculus cannot for centuries compare with the Greek theory of the continuum as regards logical rigor. But on the other hand it has widened the range of problems for it attacks from the very first the analysis of arbitrary continuous forms and processes, especially processes of motion. In our cultural sphere the passionate urge toward reality is more powerful than is the clear-sighted Greek *ratio*.

The limiting process finally won the victory, and thereby this second attempt to transfix the becoming continuum into rigid being had also failed. For the limit is an inevitable notion the importance of which is not affected by our accepting or rejecting the infinitely small. And once it has been conceived, it is seen to make the infinitely small superfluous. Infinitesimal analysis purports to derive the behavior of the finite from the behavior of the infinitely small, the latter being governed by elementary laws; so, for example, it deduces from the universal law of attraction for two material "volume elements" the magnitude of the attraction of arbitrarily shaped, extended bodies with homogeneous or non-homogeneous mass distribution. But if one does not here interpret the infinitely small "potentially" in the sense of the limiting process, the one has nothing to do with the other, their behaviors, the one in the domain of the finite and the other in the domain of the infinitely small, become entirely independent of one another, and the connecting link is broken. Here Eudoxos' view was doubtless correct.

At first it may seem as though this victory of the limiting process ultimately realized Aristotle's doctrine that the infinite exists only *dynamei*, in potentiality, in the state of becoming and ceasing to be, but not *energeiai* [in actuality]. This is far from true! The efforts to establish the foundation of analysis in the nineteenth century from Cauchy to Weierstrass, which start out from the limit notion, result in a new, powerful attempt to overcome the dynamics

of the infinite in favor of static concepts: the theory of sets. The individual
convergent sequence, such as, for example, the sequence of the partial sums of
the Leibniz series $\frac{1}{1} - \frac{1}{3} + \frac{1}{5} - \frac{1}{7} + \cdots$, which converges toward $\frac{\pi}{4}$, does not
develop according to a lawless process to which we have to entrust ourselves
blindly in order to find out what it produces from step to step; instead it is fixed
once and for all by a definite law which associates with every natural number
n a corresponding approximate value of the series, the nth partial sum. But
law is a static concept. If we ask what is meant by the convergence of the
sequence of points $P_1, P_2, \ldots, P_n, \ldots$ toward the point P, analysis supplies
the answer: it means that for every positive fraction ϵ there exists a natural
number N such that the distance PP_n is smaller than ϵ for all indices $n \geq N$.
The dynamics of the transition to the limit is here reflected in a static relation
between the sequence $\{P_n\}$ and the point P, a relation which indeed can only
be formulated by an unrestricted use of the terms "there exists" and "all" in
connection with the sequence of natural numbers. This standpoint characterizes what was previously called the fourth stage of arithmetic. Consider the
definition "n is an even or an odd number according as there exists or does
not exist a number x for which $n = 2x$." For one who accepts this definition
with its appeal to the infinite totality of numbers x as having a meaning, the
sequence of numbers open into infinity has transformed itself into a closed
aggregate of objects existing in themselves, a realm of absolute existence which
"is not of this world," and of which the eye of our consciousness perceives
but reflected gleams. In this absolute realm the *tertium non datur* [the law
of the excluded middle] is valid with regard to every property P predicable
of numbers. This implies the alternative: either there exists a number of the
property P, or all numbers have the opposite property non-P. But this could
be decided in all circumstances only if one could examine the entire sequence
of numbers with regard to the property P, which contradicts the nature of the
infinite. We are therefore prohibited from interpreting an existential proposition as the completed logical sum:[20] "1 has the property P, or 2, or 3, or ...
in infinitum," and from interpreting the general proposition as the logical
product: "1 has the property non-P, and 2, and 3, and ... in infinitum." But
then the general proposition can only be understood hypothetically as asserting
something only if a definite number is given, and it is therefore not deniable.
The existential proposition then has a meaning only with regard to a definite
example: this definite number constructed in such and such a way has the
property P. Existential absolutism disregards such difficulties which spring

from the nature of the infinite, and accepts these propositions as ordinary judgments capable of negation and opposing one another in the *tertium non datur*.

The theory of sets, in its endeavor to establish a foundation for analysis, has to go much further: it applies the terms "there exists" and "all" without limitation also to the possible sequences and sets of natural numbers—implying that such propositions refer to an actual state of affairs which lies decided within the things themselves as by "Yes" or "No," even if mathematical investigation may succeed only through a lucky chance in transforming this latent answer into an articulate one. We speak of the set of all even numbers, the set of all prime numbers. A set is thus always described as follows: the set of all numbers of such and such a property; the set is considered as given if a definite criterion decides which elements belong to it and which do not. But if the question arises whether among all possible sets and sequences a set of such and such a kind exists, one can hardly help feeling that through the determination of the "accessible" sets and sequences by laws, a chaotic abundance of possibilities, of "lawless" sets, "arbitrarily thrown together" is lost, and that thereby the clear alternative "does there or does there not exist?" is confused. The theory of sets unhesitatingly makes use of such alternatives in the criteria which it sets up to decide whether a point or a number belongs to a set or not. One can see that it thereby becomes involved in fatal logical circles. It is true that so far no actual contradictions in analysis proper have resulted; we do not completely understand this fact at present. G. Cantor, however, cast off all bonds by operating absolutely freely with the notion of sets, and in particular by permitting that from every set the set of all its subsets may be formed. And only here at the very outskirts of the theory of sets does one incur actual contradictions. Their root, however, must be traced to the bold act mathematics has performed from its very start: treating a field of constructive possibilities like a closed ensemble of objects existing in themselves.

The criticism of H. Poincaré, B. Russell, and mainly Brouwer during the last thirty years has gradually opened our eyes to the untenable logical position from which the method of the theory of sets started out. To my mind, there can no longer be any doubt that this third attempt has also failed—failed in the sense in which it was undertaken. I shall therefore consider now the fourth and last attempt. As D. Hilbert recognized, mathematics may be saved without diminishing its classical content only by a radically new interpretation through a formalization which, in principle, transforms it from a system of knowledge into a game with signs and formulas played according to fixed rules. By extending the symbolic representation customary in mathematics to the logical operations "and," "or," "there exists," and so forth, every mathematical proposition is transformed into a meaningless formula composed of signs, and mathematics itself into a game of formulas regulated by certain conventions—comparable indeed to the game of chess. To the men in the chess game corresponds a limited—or unlimited—supply of signs in mathematics; to an arbitrary position of the men on the board, the combination of the signs into a formula. One formula or several formulas are considered as axioms; their counterpart in the game of chess is the prescribed position of the men at the beginning of the game. And as in chess a new position is produced from the preceding one by a move that has to satisfy certain rules, so in the case of mathematics, formal rules of conclusion are set down according to which new formulas can be obtained, i.e., "deduced" from given ones. Certain formulas of intuitively described characteristics are branded as contradictions; in the chess game we may consider as a "contradiction" any position in which, for example, more than eight white pawns occur. So far all is game and not cognition. But in "metamathematics," as Hilbert says, the game itself becomes the object of cognition: we want to know that a contradiction can never occur as the terminal formula of a proof. This consistency of classical analysis and not its truth is what Hilbert wishes to insure; the truth we have renounced, of course, by abandoning its interpretation as a system of significant propositions. Analogously it is no longer game but cognition, when one proves that in a correctly played chess game more than eight white pawns are impossible. This is done in the following way. At the beginning there are eight pawns; by a move corresponding to the rules, the number of pawns can never be increased; ergo. . . . This ergo stands for a conclusion by complete induction which follows the moves of the given game step by step to the final position. Hilbert needs significative thinking only to obtain this one cognition; his consistency proof, in principle, is conducted like the one just carried through for the chess game, although of course it is much more complicated. It is clear that in these considerations the limitations set by Brouwer for significative thinking are respected.

From this formalistic standpoint the question as to a deeper reason for the adopted axioms and rules of operation is as meaningless as it is in the chess game. It even remains obscure why it is of concern to us that the game shall be consistent. All objections are obviated, since nothing is asserted; rejection could only take the form of the declaration: I will not join in the game. If mathematics would seriously retire to this status of pure game for the sake of its safety, it would no longer be a determining factor in the history of the mind. De facto it has not performed this abdication and will not perform it. Hence we must after all attempt to reassign to mathematics some function in the service of knowledge. Hilbert expresses himself somewhat obscurely to the effect that the infinite plays the part of an idea in the Kantian sense, supplementing the concrete in the sense of totality. If my understanding is correct, this function is analogous to the act by which I supplement the objects actually given to me in my consciousness to form the totality of an objective world, which also comprises many things that are not immediately before me.

The scientific formulation of this objective conception of the world takes place in physics, which employs mathematics as a means of construction. But the situation that prevails in theoretical physics in no way corresponds to Brouwer's ideal of a science, to his postulate that every proposition shall have its individual meaning, and that this meaning shall be capable of intuitive display. On the contrary, the propositions and laws of physics taken individually do not have a content which can be verified experimentally; it is only the theoretical system as a whole which can be confronted with experience. What is achieved is not intuitive cognition of an individual or general state of facts, and a description which faithfully portrays the given conditions, but theoretical, purely symbolical construction of the world.

The considerations of all three lectures lead from different directions to this basic view. Taking the most primitive object of mathematics, the sequence of natural numbers, as an example, I have outlined the transition from description which merely subsumes the actually occurring numbers under descriptive characteristics and relations to the construction of a field of possibilities which is open into infinity. The latter is precisely the quantitative method of physics; it does not, for example, classify the given colors as Linnaeus classified the actually occurring plants; but it reduces them to the scale of wavelengths, that is, it embeds them in a continuum constructed according to the above described division scheme, in which every possible color must find its place. We found in the second lecture that elementary analysis must be carried so far as to establish elements varying each exclusively within a range of possibilities which can be surveyed from the start, because it originates from free construction. This is one side of the matter which I designated by the term constructive generation of a field of variation. The other side which here mainly interests us is the fact that the subsumption of the particular concrete case, the "individual," into this field does not take place on the basis of immediately recognizable characteristics, but as the result of mental or physical manipulations or reactions to be performed on it. To determine number, for example, one has to apply the process of counting; to determine the mass of a body one has to allow it to react with other bodies and apply the law of momentum to the impact. But this analytical method furnishes "ideal attributes" and not concrete properties. We ascribe these ideal attributes to the objects, even if the manipulations necessary to "measure" them are not really carried out. If we indicate the distance of the sun from the earth in feet, this statement would acquire a meaning verifiable in the given state of facts only if a rigid pole on which the individual division had been marked by laying off a movable, rigid measuring rod were so applied on the earth that its end touched the sun. But this rigid pole between earth and sun does not exist, the measurement by a rigid rod is not really carried out. Geometrical statements of this kind consequently lack a meaning that can be exhibited in the given facts; the network of ideal determinations touches experiential reality only here and there, but at these points of contact ideal determination and experience must agree. Quantum theory has shown that the transformation of the results of possible reactions into properties is precarious. One may without hesitation combine two properties with each other by "and," but not so the results of two measurements, if the performance of one makes the performance of the other impossible in principle.[21]

To illustrate what the required concordance between theory and experience consists of, let us take the following example, chosen as simple as possible. We observe one single oscillation of a pendulum; let us assume that it is possible to observe its duration directly with an error of 0.1 second, so that periods of oscillation differing theoretically by less than 0.1 second are actually equal for our direct perception. There is, however, a simple means of increasing the exactness one hundredfold: one waits until 100 oscillations have taken place and divides the observed time interval by 100. But this indirect determination is dependent on an assumption, namely that all individual oscillations take the same time. This can of course be tested with an exactness of 0.1 second by direct observation. But that is not meant here. We wish instead to assert that the periods of oscillation are absolutely equal or equal with hundredfold precision. This assumption, as well as the assertion concerning the duration of an individual oscillation, is meaningless for the intuitionist who respects the limits of intuitive exactness. Still, a test of the theory is possible in a certain sense: one finds that the duration of m successive oscillations is to that of n oscillations as $m : n$, when m and n are large numbers. (For the test several series of consecutive oscillations are arbitrarily chosen.) In general the matter is as follows: through the exact laws of the theory which is taken as a basis, the quantity x to be determined is placed in functional dependence on a number of other quantities. By observing these quantities conclusions can be drawn as to the value of x, which permit us to ascertain x more precisely than is possible by its direct observation. The underlying theory is considered to hold good, if within the limits of error to be expected all indirect methods of determining x lead to the same result. But every such indirect determination, every distinction not existing for intuitive perception, is possible only on the ground of theories. These theories can only be verified by observing that when tested in all their numerical consequences, they furnish concordance within the limits of error.

It is a deep philosophical question, what "truth" or objectivity we are to assign to theoretical construction as it extends far beyond the actually given. The concordance just discussed is an indispensable requirement that every theory must satisfy. It includes, however, the consistency of the theory, so that here also we receive a rational answer to the question as to why the consistency of formalized mathematics is of importance to us: it is that part of concordance which relates only to the theory itself, the part in which the theory is not yet confronted with experience. It is the task of the mathematician to see that the theories of the concrete sciences satisfy this condition sine qua non: that they be formally definite and consistent. My opinion may be summed up as follows: if mathematics is taken by itself, one should restrict oneself with Brouwer to the intuitively cognizable truths and consider the infinite only as an open field of possibilities; nothing compels us to go farther. But in the natural sciences we are in contact with a sphere which is impervious to intuitive evidence; here cognition necessarily becomes symbolical construction. Hence we need no longer demand that when mathematics is taken into the process of theoretical construction in physics it should be possible to set apart the mathematical element as a special domain in which all judgments are intuitively certain; from this higher viewpoint which makes the whole of science appear as one unit, I consider Hilbert to be right.

In concluding I shall try to put together in a few general theses the experiences which mathematics has gained in the course of its history by an investigation of the infinite.

1. In the spiritual life of man two domains are clearly to be distinguished from one another: on one side the domain of creation (*Gestaltung*), of construction, to which the active artist, the scientist, the technician, the statesman devote themselves; on the other side the domain of reflection (*Besinnung*) which consummates itself in cognitions and which one may consider as the specific realm of the philosopher. The danger of constructive activity unguided by reflection is that it departs from meaning, goes astray, stagnates in mere routine; the danger of passive reflection is that it may lead to incomprehensible "talking about things" which paralyzes the creative power of man. What we were engaged in here was reflection. Hilbert's mathematics as well as physics belongs in the domain of constructive action; metamathematics, however, with its cognition of consistency, belongs to reflection.

2. The task of science can surely not be performed through intuitive cognition alone, since the objective sphere with which it deals is by its very nature impervious to reason. But even in pure mathematics, or in pure logic, we cannot decide the validity of a formula by means of descriptive characteristics. We must resort to action: we start out from the axioms and apply the practical rules of conclusion in arbitrarily frequent repetition and combination. In this sense one can speak of an original darkness of reason: we do not have truth, we do not perceive it if we merely open our eyes wide, but truth must be attained by action.

3. The infinite is accessible to the mind intuitively in the form of the field of possibilities open into infinity, analogous to the sequence of numbers which can be continued indefinitely; but

4. the completed, the actual infinite as a closed realm of absolute existence is not within its reach.

5. Yet the demand for totality and the metaphysical belief in reality inevitably compel the mind to represent the infinite as closed being by symbolical construction.

I take these experiences derived from the development of mathematics seriously in a philosophical sense. The mathematical tendencies which first announced themselves with Nicholas of Cusa have, as I have tried to explain, been elaborated in the course of the centuries and have reached their fulfillment. I therefore ask you to consider the content of this lecture as a more precise exposition referring back to what I said in the first lecture concerning the mathematical and theological ideas of Nicholas of Cusa. If, following his steps, we may undertake to give a theological formulation to our last three conclusions, we may say this:

We reject the thesis of the categorical finiteness of man, both in the atheistic form of obdurate finiteness which is so alluringly represented today in Germany by the Freiburg philosopher Heidegger, and in the theistic, specifically Lutheran-Protestant form, where it serves as a background for the violent drama of contrition, revelation, and grace.[22] On the contrary, mind is freedom within the limitations of existence; it is open toward the infinite. Indeed, God as the completed infinite cannot and will not be comprehended by it; neither can God penetrate into man by revelation, nor man penetrate to him by mystical perception. The completed infinite we can only represent in symbols. From this relationship every creative act of man receives its deep consecration and dignity. But only in mathematics and physics, as far as I can see, has symbolical-theoretical construction acquired sufficient solidity to be convincing for everyone whose mind is open to these sciences.

5 ▣

Mind and Nature

1934

▣   I. Subjective Elements in Sense Perception

It shall be the purpose of these lectures to trace a characteristic outline of the mathematical-physical mode of cognition. In doing this I should like to place one theme foremost: the structure of our scientific cognition of the world is decisively determined by the fact that this world does not exist in itself, but is merely encountered by us as an object in the correlative variance of subject and object. The world exists only as that met with by an ego, as one appearing to a consciousness; the consciousness in this function does not belong to the world, but stands out against the being as the sphere of vision, of meaning, of image, or however else one may call it.

Separately, the lectures are disposed as follows. First we want to make clear to ourselves from the naive realistic standpoint of the natural or of the scientif-ically educated person, to what extent our ideas of the objects of the external world are dependent on our psycho-physical organization; this shall be carried out in detail particularly in the domain of optical perceptions. In the second lecture will follow the epistemological reflection which questions the realistic standpoint in principle; and simultaneously the theory of the subjectivity of the sense qualities will be extended to the perceptual forms of space and time. With regard to scientific cognition we conclude herefrom that it does not state and describe states of affairs—"Things are so and so"—but that it constructs symbols by means of which it "represents" the world of appearances. The third lecture shows, particularly with reference to the formation of constructive notions and theories in natural science, how the field of the possible and free choice in such a field must necessarily be opposed to and placed ahead of the really occurring and the confinement to mere actuality. In order that these methodological considerations shall not remain too abstract, all that has been said will be exemplified in the fourth lecture on the Einstein relativity theory. At the end I should finally like to discuss modern quantum theory which has thrown an entirely new light on the relation of subject and object in scientific cognition.[1]

Today we shall begin by analyzing, mainly in the domain of visual and especially of color perceptions, the relationship between subject and object as it presents itself to the view of the scientifically educated realist. Visual perceptions are transmitted to us by the light. Let us follow its course from the object which irradiates, disperses, or reflects it to the retina. Our visual sense evidently does not give us full-value reality, the thing as it really is. Ordinary consciousness even is led to realize the semblant nature of the images before it by perspective and shadows; or—if these occurrences have become too commonplace to arouse astonishment—by phenomena like the rainbow and the deceptive mirror image. The word of many a poet testifies to the mental emotion which the realization of the semblant nature of that which is seen can produce, by urging upon man the anxious question as to his own reality:

Leben wir alle nur in Spiegelland?
Leben wir alle nur in Spiegellicht?[2]

Do we all live but in mirror land?
Do we all live but in mirror light?

The optical image of an object does not simply detach itself from the object and walk into my brain or present itself to my perception unchanged and without intermediary; but the light affecting my sense organ is produced by the object and propagates itself through space according to physical laws. Therefore the image seen by me by no means renders the object itself but depends not only on this object but on all accompanying physical circumstances. For example, a reflecting surface placed in the path of the light causes the rays to reach my eye in a manner in which they could reach it without such an artificial contrivance only if the object of which the reflected image gives the illusion really existed. In a similar fashion, the refraction of the sunlight in the water drops of the clouds illusorily displays before my eyes the rainbow. The color qualities with which the objects of the external world vest themselves for me depend essentially on the illumination. Is cinnabar red? Yes, it appears so in white light, but in greenish light it appears almost black. Which color appearance is the correct one? That is a foolish question; the first appearance merely distinguishes itself by corresponding to familiar, normal circumstances, as the white sunlight is our natural source of light. The physicist explains the state of affairs as follows: Of all spectral colors cinnabar, due to the constitution of its molecules, reflects almost exclusively red, while it absorbs the other colors to the greatest part. The spectral composition of the light reflected by the cinnabar, besides depending on the constitution of the cinnabar, naturally depends on the spectral composition of the light falling on it. The objective constant property of cinnabar which corresponds to the red color perceived under normal circumstances thus, theoretically speaking, lies in its molecular constitution; or, in a more phenomenological wording, in the law according to which the spectral composition of the light reflected by the cinnabar arises from the spectral composition of the incident light.

We abstract from our example the general law of psycho-physical near action: No sense, the sense of vision as little as any other, really reaches out into the distance; what I see is determined only by the condition of the optical field in its zone of tangency with the body of my senses (that is, for the visual sense the retina).

The optical "image" caught by the retina is to a certain degree similar in its geometric form to the object—or rather to that view of the object which geometrical optics traces according to the laws of perspectivity. We owe this fact to a fortunate circumstance which is, however, in itself physically accidental: that the wavelength of the light serving us for perception is small as compared to the dimensions which are vitally important for us about the external objects. (For this reason it does not matter that details of the image which are of the order of magnitude of the wavelength are blurred by the spreading of the sense irritation from the directly affected spot of the retina over the neighborhood: the precision of the image finds its natural limitation in the structure of the sense apparatus receiving the impressions. This only by the way.) But when dimensions of the object which have the order of magnitude of a wavelength are drawn out in the image to perceptible distances, the geometric similarity between the object and the image is completely lost. If a slit whose thickness amounts to only a few wavelengths is illuminated, the optical picture, due to the diffraction of the light, consists of a whole series of parallel bands. If a crystal composed of atoms regularly arranged in a latticelike fashion at distances of the order of magnitude 10^{-8} cm = 1 Å is irradiated with X-rays whose wavelength is of the same order of magnitude there arise on the photographic plate the famous Laue interference figures which surely in no sense give a similar geometric image of the atoms in their mutual arrangement. How different the world would appear to us, and possibly how much more difficult it would be to find one's way about in it on the ground of optical images, if our eye were susceptible to other wavelengths!

I suppose I should add a few words about the nature of light in order that these considerations may become fully comprehensible. According to the teachings of physics, light consists of an oscillating process which propagates itself with the enormous velocity $c = 186,284$ miles per second. What oscillates is not so important for us. (One used to assume that there was an oscillating substance, the light ether, similarly as sound waves have the substantial air for their medium. But this hypothesis led to more and more unbearable contradictions. Today we know that what oscillates is a weak electromagnetic field. If this is only a catchword for you, let the matter rest for the moment.) A simple oscillation is characterized by its *frequency* v, the number of oscillations per second. This number simultaneously fixes the wavelength λ of the propagated wave according to the law $\lambda = c/v$. The frequency or wavelength manifests itself perceptually in the color. But our eye is susceptible to light only in a certain frequency range, the perceptible spectrum from extreme red to violet extends from about $\lambda = 7500$ Å down to waves of length 4000 Å; that is, less than an octave—if in the sense of analogy I make use of the language of music which is coined for sound waves.[3] The waves of greater or smaller wavelength are by no means unknown to us; we can demonstrate them by their physical effects and show that they behave like light rays in every respect. Towards the red end follow the infra-red heat rays up to the long waves of several cm to km which are made use of in wireless telegraphy; towards the violet side we enter the ultra-violet range and get down to wavelengths of the order of magnitude 1 Å, the wavelengths of the X-rays; the "cosmic" rays of unknown origin which penetrate to us from interstellar space are even of a still shorter wavelength. Unfortunately the retina is a receptive apparatus for the wavelengths of the ether which responds only to a rather narrow interval. The photographic plate is much better in this respect. The circumstance that our eye should respond precisely to wavelengths around 6000 Å is doubtlessly connected with the fact that the natural light in which the living beings on the earth see things, the

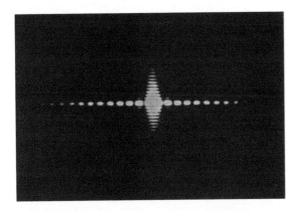

Figure 5.1 Diffraction pattern of a small slit. (Width 1/4 mm; source of light: circular diaphragm illuminated from behind. Photograph by A. Köhler. From M. Born, *Optik*, Berlin, 1933.)

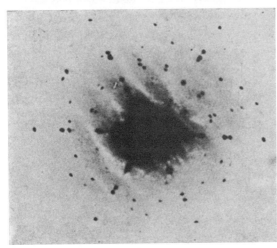

Figure 5.2 Laue interference pattern of a crystal. From a photograph by Dr. G. L. Locher, Bartol Research Foundation.

sunlight (when it is spectrally decomposed) has its maximum of intensity in this neighborhood. Speculations as to how this particular adaptation of the living beings to their surroundings came about must be left to the biologists.[4]

With our last considerations we have passed unawares from the influence which the accompanying physical circumstances exert on our sense perceptions to an influence which penetrates much more deeply and destroys every similarity between the original and the image, that, namely, of our psycho-physical organization. The retina, as we have seen, suppresses by far the greatest part of the "lights" radiating through the universe. But it distorts and reduces in an excessive manner even the manifold of those which it exhibits. Before we describe this in more detail a general principle of sense physiology shall be advanced.

The qualities that we attribute to the objects of the external world correspond to our five senses: so far they are even completely determined by the physiological structure of our sense body. One and the same physical agent affects different senses. The radiation coming from the sun gives light and heat; whatever intercepts or deflects the ray suspends both effects. The physicists hesitated for a long time and examined and refuted all possible objections before they admitted the identity of light and heat rays, the essential difference between which seemed to manifest itself in the sensation of light and heat. Conversely as regards the individual sense, it can be excited only to its specific sensations, but it is excited to that by all kinds of causes. The eye is excited to light sensations not only by physical light which falls into it, but also by electrical currents, by an impact or a pressure on the eyeball, by the introduction of narcotic poison in the optic nerves, in the case of operations by the irritation in the wound on the stump of the optic nerve. Light is the stimulus adequate to the eye only in the sense that the eye is well protected by its position in the body against all influences other than those of light and is infinitely more susceptible to light than the other sense organs. The facts here presented were summed up in their fundamental importance first by the physiologist Johannes Müller at the beginning of the nineteenth century in his law of the "specific energies of the senses." Incidentally, the specific character of the sense energies is surely not to be localized in the nerves but only in the terminal organ in the brain. For as experiences with operations have taught, nerves of different sense domains can replace one another functionally. They behave like telegraph wires which by means of the electric current now ring a bell, now set a writing telegraph in motion, now produce a chemical decomposition, and so on.

But now let the retina be excited by its adequate stimulus, light; and let us study a little more precisely how the retina represents the manifold of objective physical colors by the process which the light excitation releases in it. Monochromatic light is completely described as to its quality by the wavelength, because its oscillation law with regard to time and its wave structure have a definite simple mathematical form which is given by the function sine or cosine. Every physical effect of such light is completely determined by the wavelength together with the intensity. To monochromatic light corresponds in the acoustic domain the simple tone. Out of different kinds of monochromatic light composite light may be mixed, just as tones combine to a composite sound. This takes place by superposing simple oscillations of different frequency with definite intensities. The simple color qualities form a one-dimensional manifold, since within it the single individual can be fixed by one continuously variable measuring number, the wavelength. The composite color qualities, however, form a manifold of infinitely many dimensions from the physical point of view. For the complete description of a compound color requires the indication with which intensity J_λ each of the infinitely many possible wavelengths λ is represented; so that it involves infinitely many independently variable quantities J_λ. In contrast hereto—what dearth in the domain of visually perceived colors! As Newton already made evident by his color-disc, they form only a two-dimensional manifold. Newton imagines the simple sequence of saturated color hues from red through the whole spectrum to violet applied in some scale on the circumference of the color-disc—the remaining gap being filled by purple which does not occur in the spectrum.[5] The second variable is the degree of saturation: in the center of the disc stands pure white, on the radius leading from a point of the circumference to the center, one passes through the various degrees of saturation produced out of the color located at the circumference by an ever-increasing admixture of white.

This discrepancy between the abundance of physical "color chords" and the dearth of the visually perceived colors must be explained by the fact that very many physically distinct colors release the same process in the retina and consequently produce the same color sensation. By parallel projection of space on to a plane, all space points lying on a projecting ray are made to coincide in the same point of the plane; similarly this process performs a kind of projection of the domain of physical colors with its infinite number of dimensions on to the two-dimensional domain of perceived colors whereby it causes many physically distinct colors to coincide. In this respect the eye is much coarser than the ear. For by the so-called "timbre" of a compound tone, the ear is very well capable of distinguishing the various proportions in which the fundamental tone and its harmonics mix; it is capable of hearing the separate tones in a chord. The eye again is extremely delicate in the perception of forms, of spatial differences. This is apparently the purpose for which it was designated in the first place by nature; of the wonderfully manifold play of the light oscillations and their superposition in space it gives us only a very feeble reproduction in the perceived colors.

It seems useful to me to develop a little more precisely the "geometry" valid in the two-dimensional manifold of perceived colors. For one can do mathematics also in the domain of these colors. The fundamental operation which can be performed upon them is *mixing*: one lets colored lights combine with one another in space (that is of course something different from the mixing of color substances on the palette of the painter). It turns out that one can represent the various colors by the points of a plane in such a manner that the colors originating by mixture from two arbitrary colors A and B, in the geometric image cover the straight line segment AB. One is furthermore tempted to arrange this representation so that when the fundamental colors A, B are mixed with the relative intensities i_A, i_B a color P results, which in the geometric picture divides the segment AB like the center of gravity of two masses of magnitude i_A, i_B applied at A and B, that is, so that

$$PB : PA = i_A : i_B.$$

It does not, however, have an intuitive meaning to compare two lights of different colors with regard to their intensity; such a comparison is possible only for lights of the same color. Consequently we arbitrarily choose two intensities i_A° and i_B° of the colors A and B which we use as units and call the color produced by mixture in this proportion: U. Now we can indicate every intensity i_A of the color A by means of the number $i_A' = i_A/i_A^\circ$ (using i_A° as unit), similarly every intensity $i_B' = i_B/i_B^\circ$ of the color B. The color P produced by mixing the two fundamental colors with the respective intensities i_A and i_B will then have to assume such a position in the geometrical picture that the anharmonic ratio $\frac{PB}{UB} : \frac{PA}{UA}$ agrees with the anharmonic ratio i_A'/i_B', more precisely $\frac{i_A}{i_A^\circ} : \frac{i_B}{i_B^\circ}$. The mathematician says that for the perceived colors and their mixture two-dimensional projective geometry holds. Namely, the fundamental relationship of projective geometry is that three points lie on a straight line, and in this geometry only anharmonic ratios of four segments like $AU, BU; AP, BP$ have an objective significance—not already ratios of two segments as in ordinary metric geometry. It is true, to be quite precise, that in color space the mixture of colors constitutes the notion not of the straight line AB but that of the straight line segment AB. The colors thus form a convex region of the complete projective plane; for it is exactly the characteristic property of a convex domain that simultaneously with two points A, B, it always contains all the points of the segment joining A and B. Thus the unambiguous result of experience is this: The perceived color qualities P can be represented continuously by the points P' of a convex region of the projective plane $P \rightarrow P'$ in such a fashion that a color P produced by a mixture of A and B is represented by a point P' which lies on the segment $A'B'$, and that

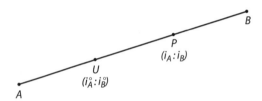

Figure 5.3 Mixing colors. (i°_A, i°_B and i_A, i_B are intensities of the colors A, B; by mixing them in these proportions the colors U and P on the "segment" AB arise. The relative intensities i_A/i°_A, i_B/i°_B are numbers. The anharmonic ratio $i_A/i^{\circ}_A : i_B/i^{\circ}_B$ determines the position of P with respect to the "coordinate system" AB, U.)

when I combine A and B to the two compound colors U and P, the anharmonic ratio of the partial segments determined on $A'B'$ by the corresponding points U', P' is equal to the anharmonic ratio of the intensities with which A and B are combined. The same laws hold for the colors and their mixture as hold in a convex region of the projective plane for the points and the segments joining them. All theorems of projective geometry as it applies to such a region have their immediate interpretation and validity in the domain of perceived colors.

Newton's mapping on the color-disc rendered only the manifold and the continuous connection of the colors correctly in the geometric image, but it was not yet a representation preserving projective relations as we now demand it. If we construct a mapping satisfying this new requirement, how does the convex region look which is covered by the image points of the perceived colors? According to the teachings of Young, Maxwell, and Helmholtz, they form (approximately) a rectilinear triangle, in the corners R, G, V of which there stands respectively a certain shade of red, green, violet. Pure white is represented by a point U (to be chosen at random) in the interior of the triangle. The mapping of the colors on the points of the triangle is thereby fully determined. The projective geometer is familiar with such a triangle consisting of three fundamental points R, G, V with a "unit point" U chosen in the interior as a coordinate system relatively to which he can fix every point of the plane in a projective manner by numbers. Young even conceived a physiological theory of color perception which is to explain how the color triangle originates. Three chemical fundamental processes, the red-, the green-, and the violet-process take place on the retina; they are excited with different intensity by the incident light. Pure red, green, and violet always excite only one of these processes. The intensity relation with which the three processes are discharged determines in general the quality of the perceived color. Later more exact observations have led to another theory, that of Hering. Here there are three chemical processes each of which can be discharged in either one of two directions. The first process when it is discharged in the one sense generates the sensation of red, when in the other sense, that of green. The second is the yellow-blue process, the third the black-white process. It must further more be assumed that every light excitation discharges the black-white process, let us say, with at least the same intensity as the two other actual color processes. (What this round-about formulation really means, is the fact that a certain restriction has to be imposed on the intensities of the two actual color processes in terms of the intensity of the black-white process.) As far as I know, one has not yet succeeded in establishing the physiological existence of these processes or of their media on the retina.[6]

With regard to the ear we are in a more fortunate position. The separation of the sound excitation into tones is performed already in the ear, before the acoustic excitation is conducted to the nerve fibers; this is due to the so-called organ of Corti, a kind of harp, the numerous strings of which respond to the various frequencies of the simple tones contained in the compound sound. From the entirely different mode of construction of the receptive apparatus in the ear and in the eye, one understands fully physiologically why in these two sense-domains the transformation of the incident excitation into perception leads to such utterly different results. We do not want to enter any further into the physiology of visual perceptions. For anyone wishing to obtain more precise information on this subject Helmholtz' classical works, *Physiological Optics* and *The Theory of Sound Sensations*, are still an appropriate guide. As regards a short summary, I refer in addition to the lectures on "Recent Progress in the Theory of Vision" which are contained in his *Gesammelte Vorträge und Reden* (Popular Scientific Lectures). Let us just keep in mind that this impoverishing projection of the physical color domain with its infinite number of dimensions on the two-dimensional manifold of perceived colors takes place already in the receptive apparatus, the retina; it does not yet touch the secret of the relationship between reality and perception, but is merely a peripheral physiological process that can be understood purely physico-chemically.

Here we stop for a moment. For the question forces itself upon us: why is physics not content with this domain of perceived colors which has only two dimensions, what urges it to put oscillations of the ether or something similar in their place? After all, from our visual perceptions we know nothing about the oscillations of the ether; what we are given are precisely only these colors, the way we encounter them in our perception. Answer: Two light rays which cause the same impression to the eye are in general distinct in all their remaining physical and chemical effects. If, for example, one illuminates one and the same colored surface with two lights which visually appear as the same white, the illuminated surface usually looks quite different in both cases. Red and green-blue together give white light, equally light brown together with violet. But the first light produces a dark hue on the photographic plate, the second a very light one. If one sends two lights which visually appear as the same white through one and the same prism, the intensity distribution in the spectrum arising behind the prism is different in both cases. Therefore physics cannot declare two lights which are perceptually alike to be really alike, or else it would be involved in a conflict with its dominating principle: *equal causes under equal circumstances produce equal effects*. Perceptual equality therefore appears to physics only as a somewhat accidental equality of the reactions which physically distinct agencies produce in the retina. The accidental equality of the reaction rests upon the particular nature of this receptive apparatus; on the photographic plate entirely different lights produce equal effects.

Incidentally, all our descriptions hold only for the normal human eye. Considerable modifications take place for the red-green blind who are not at all infrequent among us. From experiments in the training of bees it has become apparent that bees are susceptible to ultra-violet light: they "fly for ultra-violet."[7]

We sum up the results of our hasty investigation with Helmholtz (see his second lecture on the theory of vision): "Of the agreement between the quality of external light and the quality of sensation there remains only one point which at first sight may appear meager enough but which is perfectly sufficient for an innumerable quantity of most useful applications: Equal light under equal circumstances produces equal color sensations. Light which under equal circumstances produces different color sensations is different." Lambert had already formulated the axiom similarly in his *Photometria*, 1760: "An apparition is the same as often as the same eye is affected in the same manner." Helmholtz explains this further by the fertile and elucidating distinction between image and sign. I shall quote literally a few sentences from the lecture just mentioned.

If two fields correspond in this manner (that the same thing in the one field ever and again corresponds to the same thing on the other side), the one is a token or a sign for the other. It seems to me that the source of numberless mistakes and false theories has lain in the fact that hitherto one has not separated the notion of a sign and the notion of an image with sufficient care in the theory of perceptions.

In an image the representation must be of the same kind as that which is represented; it is an image only so far as it is like in kind. A statue is an image of a person so far as it portrays his bodily form by its own bodily form. Even if it is executed on a reduced scale, a space magnitude is always represented by a space magnitude. A picture is an image of the original, partly because it imitates the colors of the latter by similar colors, partly because it imitates its spatial relations by similar spatial relations (by means of perspective projection).

The nerve excitations in our brain and the conceptions in our consciousness can be images of the occurrences in the external world so far as the former reproduce by their sequence in time the sequence in time of the latter, in so far as they represent equality of objects by equality of signs, and therefore also lawful order by lawful order. This is evidently sufficient for the task of our intelligence which in the colorful variety of the world must detect that which remains constant and sum it up as a notion or a law.

Then a little later:

One may not allow oneself to be led astray into confounding the notions of appearance and semblance. The bodily colors are the appearance of certain objective differences in the constitution of the bodies; they are thus also in the scientific view by no means void semblance, even though the manner in which they appear depends principally on the constitution of our nervous apparatus. A deceptive semblance occurs only when the normal mode of appearance of one object is interchanged with that of another. But this by no means takes place in the perception of colors; there exists no other mode of appearance of the colors which we could designate to be the normal one as compared to that in the eye.

At another place Helmholtz formulates the same recognition generally like this: "A difference in the perceptions urging themselves upon us is always founded in a difference of the real conditions."[8] But in these real conditions the body of the observer making the perception is naturally included as a physical thing. And he states this principle of the "empiristic view":

> The sensations are signs for our consciousness of which to learn to understand the meaning is left to our intelligence. . . . When we have learnt to read these symbols correctly we shall be able with their help to adjust our actions so as to bring about the desired result, that is so that the expected new sensations will arise. A different comparison of conceptions and things not only does not exist in reality—all schools agree in that point—but a different kind of comparison is absolutely inconceivable and has absolutely no meaning.[9]

Mathematics has introduced the name isomorphic representation for the relation which according to Helmholtz exists between objects and their signs. I should like to carry out the precise explanation of this notion with regard to the correspondence between the points of the projective plane and the color qualities, which I referred to above. On the one side, we have a manifold Σ_1 of objects —the points of a convex section of the projective plane which are bound up with one another by certain fundamental relations $R, R', \ldots$; here, besides the continuous connection of the points, it is only the one fundamental relation: "The point C lies on the segment AB." In projective geometry no notions occur except such as are defined logically on this basis. On the other side, there is given a second system Σ_2 of objects—the manifold of colors—within which certain relations $R, R', \ldots$ prevail which shall be associated with those of the first domain of objects by equal names, although of course they have an entirely different intuitive content. Besides the continuous connection, it is here the fundamental relation: "C arises by mixture from A and B"; let us therefore express it somewhat strangely by the same words which we used in projective geometry: "The color C lies on the segment joining the colors A and B." If now the elements of the second system Σ_2 are made to correspond to the elements of the first system Σ_1 in such a way, that to elements in Σ_1 for which the relation R, or R', or . . . holds, there always correspond elements in Σ_2 for which the homonymous relation is satisfied, then the two domains of objects are isomorphically represented on one another. In this sense the projective plane and the color continuum are isomorphic with one another. Every theorem which is correct in the one system Σ_1 is transferred unchanged to the other Σ_2. A science can never determine its subject-matter except up to an isomorphic representation. The idea of isomorphism indicates the self-understood, insurmountable barrier of knowledge. It follows that toward the "nature" of its objects science maintains complete indifference. This—for example what distinguishes the colors from the points of the projective plane—one can only *know* in immediate alive intuition. But intuition is not blissful rest in itself from which it may never step forth, but it urges on toward the variance and venture of *cognition*. It is, however, fond dreaming to expect that by cognition a deeper nature than that which lies open to intuition should be revealed—to intuition.

These somewhat anticipatory speculations were brought about by the lack of similarity which prevails between the physical colors and the processes excited by them on the retina, and by the Helmholtz "sign theory" of sensations, which we abstracted from it. The processes on the retina produce excitations which are conducted to the brain in the optic nerves, maybe in the form of electric currents. Even here we are still in the real sphere. But between the physical processes which are released in the terminal organ of the nervous conductors in the central brain and the image which thereupon appears to the perceiving subject, there gapes a hiatus, an abyss which no realistic conception of the world can span. It is the transition from the world of being to the world of the appearing image or of consciousness. Here we touch the enigmatic twofold nature of the ego, namely that I am both: on the one hand a real individual which performs real psychical acts, the dark, striving and erring human being that is cast out into the world and its individual fate; on the other hand light which beholds itself, intuitive vision, in whose consciousness that is pregnant with images and that endows with meaning, the world opens up. Only in this "meeting" of consciousness and being both exist, the world and I.

From the physical-physiological considerations of this first lecture we herewith turn in the second to critical epistemological reflection.

II. World and Consciousness

The world does not exist independently but only for a consciousness. We realize by epistemological reflection that a quality like color can be given only in consciousness, in sensation; that it has no meaning at all to assign color to a thing in itself as a property detached from consciousness. This insight by which naive realism is in principle overcome constitutes the beginning of all our philosophizing. It opposes the epistemological standpoint of idealism to that of realism in teaching us to consider the immediate data of consciousness as primary, the world of objects as secondary.

The doctrine of the subjectivity of sense-qualities has been connected with the progress of science since Democritus (460–360 B.C.) laid down the principle: "Sweet and bitter, cold and warm as well as the colors, all these things exist but in opinion and not in reality (*nomōi, ou physei*); what really exists are unchangeable particles, atoms, and their motions in empty space." I will quote another fragment by Democritus that has come down to us: "We really do not perceive anything which is not delusive; we receive merely impressions that change corresponding to the varying states of our body and the images which come crowding in upon us." Plato in the dialogue *Theaetetus* expresses himself in a similar way. Galileo may be mentioned as another witness:

> White or red, bitter or sweet, sounding or silent, sweet-smelling or evil-smelling are names for certain effects upon the sense-organs; they can no more be ascribed to the external objects than can the tickling or the pain caused sometimes by touching such objects.

Locke, among the philosophers of the Age of Enlightenment, discusses the subjective character of sense-qualities in fullest detail; his work *On Human Understanding* deals with this question in book 2, chapter VIII. His is the classical distinction of primary and secondary qualities. Primary ones are extension, shape (i.e., all geometrical properties), motion and solidity; secondary qualities are those

> which in truth are nothing in the objects themselves, but powers to produce various sensations in us by their primary qualities, i.e., by the bulk, figure, texture and motion of their insensible parts, as colors, sounds, tastes, etc.

Locke's conviction is this:

> The ideas of primary qualities of bodies are resemblances of them, and their patterns do really exist in the bodies themselves; but the ideas produced in us by these secondary qualities have no resemblance of them at all. . . . What is sweet, blue or warm in idea, is but the certain bulk, figure, and motion of the insensible parts in the bodies themselves, which we call so.

I shall read to you the beginning of §18 of the same chapter:

> A piece of manna of a sensible bulk, is able to produce in us the idea of a round or square figure; and by being removed from one place to another, the idea of motion. This idea of motion represents it, as it really is in the manna moving; a circle or square are the same, whether in idea or existence; in the mind or in the manna. And this, both motion and figure are really in the manna, whether we take notice of them or not. This everybody is ready to agree to. Besides, manna by the bulk, figure, texture and motion of its parts has a power to produce the sensation of sickness, or sometimes of acute pains or gripings in us. That these ideas of sickness and pain are not in the manna, but effects of its operations on us, and are nowhere when we feel them not: this also everyone readily agrees to. And yet men are hardly to be brought to think, that sweetness and whiteness are not really in manna; which are but the effects of the operations of manna, by the motion, size and figure of its particles on the eyes and palate; as the pain and sickness caused by manna are confessedly nothing but the effects of its operations on the stomach and guts, by the size, motion and figure of its insensible parts.

So far Locke.

May we finally quote Hobbes. To prove his theory of the "Unreality of Consciousness" he refers to the phenomenon of reflection in a mirror, just as we did at the beginning of the first lecture. He speaks of "the great deception of sense, which is also to be by sense corrected," and he formulates his thesis in these words:

1. that the subject wherein color and image are inherent is not the object or thing seen;
2. that there is nothing without us (really) which we call an image or color;
3. that the said image or color is but an apparition unto us of the motion, agitation or alteration which the object worketh in the brain or spirit.

The idea of the merely subjective, immanent nature of sense qualities, as we have seen, always occurred in history woven together with the scientific doctrine about the real generation of visual and other sense perceptions—a topic which we dealt with earlier during the first lecture. Locke's standpoint in distinguishing primary and secondary qualities corresponds to the physics of Galileo, Newton, and Huyghens; for here all occurrences in the world are constructed as intuitively conceived motions of particles in intuitive space. Hence an absolute Euclidean space is needed as a standing medium into which the orbits of motion are traced. One can hardly go amiss by maintaining that the philosophical doctrine was abstracted from or developed in close connection with the rise of this physics.

Leibniz seems to have been the first to push forward to a more radical conception. "Concerning the bodies," he says, "I am able to prove that not only light, color, heat and the like, but motion, shape and extension too are mere apparent qualities." Later on Berkeley and Hume took somewhat the same position. This recognition that transcends Locke found its classical expression in Kant's *Transcendental Aesthetic*; according to him, space and time are forms of our intuition:

> That in which sensations are merely arranged, and by which they are susceptible of assuming a certain form, cannot be itself sensation; hence, indeed, the matter of all phenomena is given to us only a posteriori (namely by means of sensations), but the form of them must lie ready a priori within our mind and therefore must be capable of being considered independently of all sensations.

It is not without interest to follow the historical development in the course of which it is finally recognized that not even space and time may be attributed to the objective world, but that they are instead intuitive forms of our consciousness; and to see why this acknowledgment was so much harder to achieve than the realization of the subjectivity of sensations. One main reason is that Locke's distinction between primary and secondary qualities was backed forcefully by the procedure of science which used spatial and temporal ideas as the material for the construction of its objective world. Hobbes in his treatise *De Corpore* starts with a fictitious destruction of the whole world; mind builds it up again out of its phantasms, the data of consciousness. There follows an a priori construction that introduces space, for instance, as phantasm of pure objective being, independent of mind. When the phantasms of sense qualities appear, the direction of Hobbes' procedure is reversed; they are not to be constructed, but we find them as they are in our mind and think over how they may be explained as the result of "motions" within the external objects and our sense organs. This looks at first as if he were advancing towards Kant's idea concerning space and time, since space equally with sense quality is introduced as what he calls phantasms. But this interpretation is contradicted by the dogmatical manner in which he propounds the mechanistic theory of the world, his contention that every change reduces in reality to motion, that action can arise only by pressing and pushing, etc. Thus we read in his treatise *On Human Nature* in complete agreement with Locke's division into primary and secondary qualities:

> Whatsoever accidents or qualities our senses make us think there be in the world, they are not there, but are seeming and apparitions only; the things that really are in the world without us, are those motions by which these seemings are caused.

Descartes, in spite of his agreement with the fact that one is not allowed to ask for a resemblance between an occurrence and its perception (sound-wave and tone, for instance) any more than between a thing and its name, nevertheless maintains that the ideas concerning space have objective validity. For what reason? He answers, because in contrast to the qualities we recognize them clearly and distinctly; and whatever we know in such a way—as a fundamental principle of his epistemology claims—is true. And in order to support that principle he refers to the veracity of God, who is not bent on deceiving us. Obviously one cannot do without such a God guaranteeing truth, as soon as one has grasped the principle of idealism but wants to build up the real world out of certain elements of consciousness that for some reason or other seem particularly trustworthy. "One sees," judges Georg Büchner, the German revolutionary and playwright of the 1830 period who wrote *Danton's Death* and *Wozzeck*, "how keenly Cartesius measured out the grave of philosophy. His use of the dear God as the ladder to climb out of it is, to be sure, strange. The attempt turned out somewhat naively, and even his contemporaries did not let him get over the edge." Finally d'Alembert no longer justifies the use of spatial-temporal notions for constructing the objective world by their clarity and distinctness like Descartes, but rather by the practical success of this method. It is certainly not permissible to think of the electrons as small colored spheres if colors are waves in the ether engendered only by the play of electronic motions. This circumstance indicates readily enough that we cannot conceive of electrons as intuitively representable bodies within intuitive space.

As for space, I do not intend to expatiate on the physiologic foundation and the psychologic constitution of our spatial intuition at such a length as

I did in the case of color qualities. As far as our optical sense is concerned in that process, we may describe the essential steps as follows: First, the two-dimensional visual field of the single eye at rest; second, the eye motions as they provide a reliable criterion for the equality of two figures given in that visual field; third, the perception of spatial depth as a third dimension by means of binocular vision which constitutes three-dimensional space with the ego as center; fourth, transition to homogeneous space, where the bodily ego takes on a position on equal terms with other bodies; this is accomplished by the possibility of walking toward the distant horizon of centered space, by the free mobility of our own body in space and by the intentions of our will directed toward such motions. Not before this last step do I become capable of imagining myself as being in the position of another person, only this space may be thought of as the same for different subjects, it is a medium necessary for constructing an intersubjective world. The other sense fields join the optical one; like the field in which we localize touch impressions and the motions of limbs. I want to dwell a little longer on only two issues in this matter.

The first question: Why do we not see the whole world upside down? The direction from bottom to top is reversed on the retina as compared with the external object; in our eye all things stand on their head. Why do we see them upright? The converting effect seems to demand that all the objects themselves should stand on their head like Father William.[10] This puzzle is merely an apparent problem. An object and its image on the retina belong to the same space that imbeds the whole external world; therefore they may be compared and thus shown to have an inverted position with respect to each other. But an analogous comparison of that real space with the space of my intuition is meaningless; here we are, so to speak, in different worlds. The utmost correctness we may demand with regard to the relation between objects and those images which are their representations in my consciousness, cannot surpass isomorphism, i.e., conservation of all proportions of length. (Unluckily not even this is the case.) It is meaningless to put questions that have no significance unless object and image exist in the same space. As long as one still finds any difficulty in this problem, one has not yet overcome the one-sided realistic interpretation of experience.

The second question at stake concerns Lotze's so-called local signs; they are meant to be sensations whose qualitative variations correspond to and mark the different positions in the field of vision.[11] Such sensations were postulated by Lotze on account of his psychological principle that nothing but qualities of sensations can be directly perceived and distinguished by our soul, and in particular that the soul must build up spatial extension out of sensations. Lotze as well as Helmholtz, H. Poincaré and others took a good deal of pains with these local signs in attempting to discover them. To me, however, this principle seems a sheer prejudice. One has to acknowledge according to Kant and Fichte: I am originally endowed with the faculty of intuition as well as sensation. A thing can exist for me only in the indissoluble unity of sensation and intuition, by the fact that a continuum of quality covers a (spatial-temporal) continuum of extension. The penetration of the *what* (here-now) and the *how* is the general form of consciousness. Space as a form of my intuition can scarcely be described more suggestively than by these words of Fichte:

> Translucent penetrable space, pervious to sight and thrust, the purest image of my awareness, is not seen but intuited and in it my seeing itself is intuited. The light is not without but within me, and I myself am the light.

My statement that color by its very nature can be met with only in sensation, did not presume that color is a peculiar property inherent in sensation. We should be more correct in saying, it is not a real component of sensation itself but rather an entity pertaining to the intentional object, which arises before my consciousness in the perceiving act. Perception, to be sure, puts a perceived object in front of me; this intentionality is a decisive character of conscious acts, quite independently of the question whether this object is taken as really existing and whether that real thing actually bears those properties which are indicated by the qualities perceived. I have the perception, I live in it. I do not perceive my perception, but rather its intentional object. The perception itself is no object but an act. However, it is a fact, and a fact of momentous import for the structure of consciousness, that I am able to become introspectively aware of my perception. So to speak, I then split myself into two parts and gaze with the eyes of mind upon my own perceiving activity. The perception itself changes, by this process of reflection, into the object of a new act, of an act of presentation or introspective perception. But then again I have this new inward perception—my life is immersed in it—that refers to the first perception as its intentional object.[12]

A full act of perception is always impregnated with certain mental interpretations which perform a kind of vitalizing integrating function on the mere sense data. We come across such a function in the domain of spatial vision, for

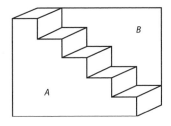

Figure 5.4 Ambiguous
perspective interpretation.

instance, when the perspective interpretation of a plane figure offered to the eye suddenly shifts.

I will show you a picture the perspective interpretation of which is ambiguous. At first glance this (fig. 5.4) is a stairway running down from left to right, with the wall A in the foreground and the wall B in the back. But if you concentrate on it you will succeed in making it change its aspect such that we now view the stairway from below; it runs up from right to left and the wall B formerly in the back, is now in the front. Try to let the two interpretations shift back and forth!

Another example: The sun is 150,000 times as bright as the full moon; consequently white paper under moonlight is darker than black velvet under sunlight; nevertheless we see that paper as white, and not as black paper, both by moonlight and by sunlight.

The resulting perception to which the sensual data thus vitalized lead, and the way in which they perform their representing function and put a concretely embodied thing before me, are both undoubtedly determined by an enormous mass of previous experiences. Consciousness reacts with an entirety that is not merely a mosaic composed of sensations; on the contrary, these so-called sensual data are a subsequent abstraction. The assertion, that they alone are actually given and the rest is derivative, is not a description that carefully pictures what is given in its full complexity, but rather a realistic theory arising from the realistic conviction that "only sensations *can* really be given." Nevertheless, one may believe in a subconscious weaving of the whole pattern out of such elements called pure sensations; Helmholtz talks of subconscious conclusions in this connection. I only want to emphasize that such hypotheses refer to psychic reality, to the sphere of being, and not of image and consciousness. They take mental phenomena as a particular domain of the real world to be studied in the manner of science by experimental methods and categories like causation, etc., as it is done most decidedly by psychoanalysis.

But let us drop these epistemological subtleties, the very source of which is the astonishing fact that this consciousness which is always I myself also occurs deprived of its immanence as the soul of an individual man in the real world! Let us turn instead toward the consequences that subjectivity of space and time bears on scientific method! Recognition of the subjective character of sense qualities induced science to discard them as building material for constructing an objective universe and to replace them by time-space notions. To be consistent, we are now urged to relinquish the ideas also of space and time as regards their serving this purpose. How can this be accomplished, how can one get rid of space and time, if one is concerned with the objective world? At first glance it seems quite impossible. But it can be done, roughly speaking, by means of replacing space-points by their coordinates. By following this device geometrical concepts are turned into arithmetical ones; for coordinates are numbers! A four-dimensional continuum in the abstract mathematical sense replaces our perceptive space and time as the medium in which physics moulds the external world. Color, which with Huyghens was "in reality" an oscillation of the ether, now appears as a mathematical law according to which the numerical values of a certain quantity called electromagnetic field strength depend on four variables or coordinates, that indicate the possible space-time positions.

Now it becomes evident that we must put this "in reality" between quotation marks. That objective world we want as a substratum under our immediate perceptions is only a symbolical construction. At least, human knowledge is unable to present it in a different form. Did Heraclitus allude to this with his aphorism handed down to us?

> The Lord whose oracle is at Delphi neither reveals nor hides, but announces by tokens.

One may admit, that only in the theory of relativity has science fully realized the fact that space and time as we know them from intuition are to be discarded as a basis of construction. The educated people of today have become thoroughly familiar with the idea that sense qualities do not show up as such in physics, but are replaced by motions in space and time, by sound or ether waves, oscillations of strings or electric currents. But the man on the street has not yet grasped this necessity of relinquishing even space and time; he can't avoid picturing those physical occurrences to himself as intuitive motions in intuitive space. This is the chief impediment that prevents the layman from a thoroughgoing understanding of the relativity theory. It is a vain hope that I should be able to master this obstacle in less than one hour, but I will try my best.

The symbolical exhibition of the universe is built up in several steps from what is immediately given; the transition from step to step is made necessary by the fact that the objects given at one step reveal themselves as manifestations of a higher reality, the reality of the next step. As a typical instance I should like to point to the manner in which the shape of a solid body is constituted as the common source of its several perspective views. This would not happen unless the point from which the view is taken could be varied and unless the standpoints actually taken present themselves as a section out of an infinite continuum of possibilities laid out within us. I shall discuss more fully the category of possibility in this connection during the next lecture. However, systematic scientific explanation will finally reverse the order: First it will erect its symbolical world by itself, without any reference, then skipping all intermediate steps try to describe which symbolical configurations lead to which data of consciousness. One instance may render clear this procedure. If the electric field strength depends on the space-time coordinates in a certain mathematically defined way, and if an eye that is awake and sees and which *I am* is present in the field, then yellow appears. The existence of the eye must, of course, be described here in the same objective symbolical way as that of the light wave (a rather complicated affair, that demands the reduction of the whole of physiology to physics). On the other hand, as we change from the transcendental sphere of objects to immanent consciousness, the further assumption that I am the living eye, is not less essential; for an apparition can only be an apparition for me. This is true no matter how violently you may protest; of course I will prevent no one from ascertaining the same fact, for himself. Another quite trivial example is afforded again by perspective: Previously the solid body constituted itself out of its perspective views; now we have to go the oppositive way and describe how a given solid body determines its two-dimensional aspect relatively to an arbitrarily given standpoint; this is done in the geometrical doctrine of perspective.

As a further instance taken from the highest stages I may mention the constitution of the concept: electric field and electric field strength. We find that in the space between charged conductors a small charged "test body" experiences a certain force $K = K(P)$, of a determined quantity and direction at every point P; the same force again and again whenever I bring the test body into the same position P. In varying the test body that here plays the part of the observer, so to speak, one realizes that the force $K(P)$ depends on it, but in such a way that $K(P)$ may be split up into two factors:

$$K(P) = e \cdot E(P).$$

The second factor, the electric field strength $[E(P)]$, is a point function independent of the state of the test body; whereas the first factor, the charge of the test body $[e]$, is a number depending only on that state, and neither on the conductors nor their position. Here we take the force K as being given to us. The outlined facts lead us to the assumption of an electric field described by the point function $E(P)$, and surrounding the conductors whether we observe the force exercised by it on a test body or not. The test body serves only to make the field observable and measurable. In this way you see the complete analogy with the case of perspective: The field E here corresponds to the object there, the test body to the observer, its charge to his position; the force that the field exercises upon the test body and which changes with the test body's charge corresponds to the two-dimensional aspect which the solid object offers to the observer, and which depends on the observer's standpoint. Now the equation $K = eE$ is no longer to be looked upon as a definition of E, but as a law of Nature (possibly needing correction) that determines the mechanical effect exercised by a given electric field E on a point charge e at a given position P. Since, according to Maxwell's theory, light is nothing else but a rapidly alternating electromagnetic field, our eye is a sense organ capable of observing certain electric fields also by other means than by their ponderomotoric effects. In a systematic representation the thing to do is to introduce an electric field strength E purely symbolically without explanations, and then to put down the laws it satisfies, together with the laws determining its ponderomotoric force. If we consider forces as the thing we can check directly, we thus link up our symbols with experience.

Analytical geometry as founded by Descartes is the device by which we eliminate intuitive space from constructive physics. After choosing a definite system of coordinates in the plane, any point can be determined and represented by its two coordinates x, y, i.e., by a pure number symbol (x, y), $(\frac{3}{2}, \frac{4}{9})$ for instance. The circumstance that several points (x, y) lie on the same straight line is expressed by a linear equation

$$ax + by + c = 0, \quad \text{like } 2x - y + 5 = 0,$$

that is satisfied by the number symbols (x, y) of all these points. Equality of two distances $P_1 P_2, P'_1 P'_2$ is expressed by equality of the corresponding numbers

$$(x_2 - x_1)^2 + (y_2 - y_1)^2, \quad (x'_2 - x'_1)^2 + (y'_2 - y'_1)^2.$$

All geometric relations thus find their arithmetic-logic representation. While the symbolic language of Newtonian physics, in terms of which it claimed to describe the whole world, was taken from intuitive space, this role is now handed over to arithmetic. The intermediate stage of spatial interpretation may well be omitted when it comes to the final systematic presentation. Only afterwards when connection is established between the symbols and our immediate conscious experiences do we have to talk about intuitive space perceptions as well as about sounds and colors—but rather on the side of consciousness than on the side of the objective world.

The propagation of a plane monochromatic light wave of frequency ν and intensity a^2 is indicated by an arithmetical expression:

$$\text{electric field strength } E = a \cdot \cos \nu \left(t - \frac{x}{c} \right)$$

containing the time and space coordinates $t; x, y, z$ as arguments (c is the constant velocity of light). We need not be so pedantic as to talk only in this "arithmetical" language, strictly avoiding all terms that refer to ideas of space and sense qualities. But, in principle, one must hold to the position that nothing of the intuitive contents and essence of these terms enters into the systematic symbolical construction of the physical world!

Heinrich Hertz—the same physicist who first showed how to produce and investigate electric waves—depicts the procedure at the introduction of his posthumous work on mechanics as follows:

> We create internal images or symbols of the external objects and we make them of such a kind, that logically necessary consequences of the symbols are always the symbols of caused consequences of the symbolized objects. A certain concordance must prevail between nature and our mind, or else this demand could not be satisfied. Experience teaches us that the demand can be satisfied and hence that such an agreement actually does exist. When we have succeeded in deriving symbols of this kind by means of previously gathered experiences we can by using them as models in a short time develop the consequences that will occur within the external world only after a long time or as

reactions to our own interferences. Thus we become capable of anticipating facts and can direct our present decisions by such knowledge. The symbols we talk of are our concepts of the objects; they agree with them in that one essential respect which is expressed by the demand mentioned above, but it is irrelevant for their purpose that they should have any further resemblance with the objects. We neither know, nor do we have means to find out whether our representations of the objects have anything in common with the objects themselves except that one fundamental relation alone.

One came to call the doctrine we are propounding here the correspondence theory of our knowledge of reality. M. Schlick is an eloquent interpreter of this doctrine in connection with relativity theory.[13] There exists a correspondence between the real world and my immediate experience, a true representation in the mathematical sense. This holds only *cum granu salis*, to be sure, as long as one really keeps to experiences that are directly given. For the single consciousness reflects only a small section of the world, and in addition, to be precise, we are only given what is given at this moment; the reliability of recollection is already a problem that transcends the purely immanent. Hence that correspondence does not hold between the one real world and the actual perceptions of an observer, but on the one side there stands the one quantitatively determined objective world as represented by our symbols, on the other side there stand the *possible* perceptions resulting from all possible objective states of an observer; things like the position and the velocity of the observer, for instance, belong to this arbitrarily variable element within the correspondence. We come here anew to that contrast of the unique fixed being of the objective world and of the freedom on the side of the observer, a contrast previously illustrated by the example of perspective. This freedom manifests itself in practical physics in the fact that the experimenter arbitrarily chooses and varies the conditions of his observations.

What compels us to refer our immediate experience to an objective symbolical world is originally, no doubt, our *belief* in the validity of recollection, in the reality of the ego, the thou and the world we live in; this belief is rooted in the last depth and is inseparably bound up with the very existence of man—that knowing, acting, caring existence that is utterly different from the existence of things! The *weltanschauliche* contrast of realism and idealism is reflected, within science, by two non-contradictory methodical principles.

Science proceeds realistically when it builds up an objective world in accordance with the demand which we previously expressed with Helmholtz that the objective configuration is to contain all the factors necessary to account for the subjective appearances: no diversity in experience that is not founded on a corresponding objective diversity. On the other hand, science concedes to idealism that this its objective world is not given, but only propounded (like a problem to be solved) and that it can be constructed only by symbols. But the fundamental thought of idealism gains prevalence most explicitly by the inversion of the above maxim: the objective picture of the world may not admit any diversity that cannot become manifest in some diversity of perception. It may actually happen that different things call forth the same impression on my mind; but this often ceases if the state of the observer is varied in all possible ways. To be sure, many physically different colors call forth exactly the same sensation of red. But by sending them through the same prism the physical difference becomes manifest in the perceivable difference of the color spectrum behind the prism. The prism, so to speak, breaks up the hidden diversity and makes it manifest for perception. But a difference that can by no means be broken up for observation, is not to be admitted. This is a maxim of construction of considerable import. Admitted that those fundamental experiences on which relativity theory is founded show the impossibility of verifying the simultaneity of events in such a way that simultaneity is independent of the state of motion of the observer and satisfies certain conditions everybody demands of this concept—this admitted, you must conclude that such a simultaneity is not present in the structure of the world. If it occurred in our previous theoretical construction of the world it must be eliminated as a superfluous element. That is what Einstein did.

With the last considerations we have already passed from the topic of this second lecture that dealt with epistemology to the topic of the next, which will be centered about scientific methodology.

III. Constructive Character of Scientific Concepts and Theories

W. Dilthey, in his essay "On the Autonomy of Thinking in the Seventeenth Century," given in the second volume of his *Collected Works*, sketches the rise of mechanics until Galileo. "Galileo came," he continues, "and with him there followed the study of an actual analysis of Nature after more than two thousand years of mere description and consideration of form in Nature, that had reached a certain summit in Copernicus' picture of the cosmos." A decisive feature of the analysis is the isolation of simple occurrences within the complexity of happenings, the dissection of the one course of events into simple and always recurring elements. Bacon already set up the formula *dissecare naturam*. "Only the mathematicians contrived to reach certainty and evidence," says Descartes, "since they started from what is easiest and simplest." The power of science is founded in no small measure on the fact that instead of designing a "System of Nature" in one draft, science has stooped with infinite patience to small isolated questions and has submitted these to an unremitting analysis. To be sure, Descartes himself still sinned heavily against his own methodological remark. Galileo's superiority to him in the field of science partly rests upon the moderation and limitation just referred to which he seriously observes and which "shows the master."[14] His investigation of the laws of falling bodies is a marvelous illustration.

This dissection of nature's course into simple always-recurring elements I shall discuss only in so far as it is related to another fundamental feature of scientific concepts, namely their indirect character. In a body we cannot perceive its inertial mass as we can its color; it can be found out only by means of reactions with other bodies. Hence in the work of dissection one does not refrain from "concealed" or, if you like, fictitious elements. In order to remain true to the principle that equal causes call forth equal reactions under equal circumstances, we interpret the simple white as a spectral compound of physical colors; and we impart such concealed differences to two colors which appear as the same white to the eye, because they can be distinguished from one another by means of a reaction, such as their passing through a prism. But light of definite spectral color and intensity proves to be physically simple, because these attributes determine entirely its behavior under all circumstances.

I think it is best if I clarify my idea by way of a concrete example of outstanding importance which pertains to the beginning of science, namely Galileo's introduction of the notion of mass. Galileo is by no means so naive as to be satisfied by the verbal, really sterile explanation: mass of a body = quantity of its matter. He looks upon mass in its dynamical function. Therefore he traces it back to the momentum or the impetus of motion. To a body that moves without being affected from outside in a straight line with constant velocity v following the law of inertia, we may ascribe a momentum I of definite magnitude and direction. The latter coincides with the direction of the

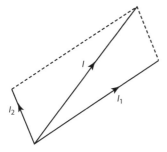

Figure 5.5   Addition of momenta.

velocity v. The inertial mass m is the factor by which v has to be multiplied to get I, or $I = m \cdot v$. But what is momentum, or, in sincere physical language how can momentum be measured? To this Galileo does not answer by a definition but by a law of nature, the law of impulse. It asserts: if several bodies react with each other, the sum of their momenta[15] after the reaction is the same as before (it is supposed that each body is isolated after as well as before the reaction against any influences from outside). By subjecting reactions, collisions of bodies, to this law one gets the means for determining empirically their relative masses.

Two bodies, for instance, that move with equal velocities in opposite directions and collide and then adhere to each other after the collision, have equal masses when neither overruns the other. We attribute a mass to the body as an inherent property, whether we actually perform such a reaction for its measurement or not, simply relying upon the *possibility* of carrying out such reactions. A very important step is accomplished hereby: After matter had been divested of all sensual qualities, science at first seemed allowed to attach only geometrical attributes to it; Descartes is quite consistent in this point; but now we see that one may derive other numerical characters of matter from its motion and from the change of motion by reactions according to certain laws. It is this method of *implicit properties* that opens the sphere of mechanical and physical concepts proper beyond geometry and kinematics. These implicit definitions are essentially bound to certain laws of nature such as the impulse law in our case. Consequently these laws appear half as expressions of experiences, half as postulates; impossible to sever the two aspects from each other. The indirect determination of quantities is possible only on the basis of a theory.

These considerations also cover the problem, how it is possible to fix a quantity much more exactly than its distinctness in sense perception allows. Let us take an oscillation of a pendulum. Direct observations may allow us to fix its duration as 1 sec with an error of ± 0.1 second. The trick by which one gets to a more precise determination is rather simple: one waits, let us say until 1000 oscillations have taken place, determines their entire duration equal to 1053.4 sec with the same error of ± 0.1 sec and infers that the simple oscillation lasts 1.0534 sec; the exactness has been increased 1000 times. This calculation, however, depends on a theoretical assumption: namely that each oscillation takes the same time. This assumption, as well as the indirectly derived assertion concerning the duration of an individual oscillation, is meaningless to the intuitionist who respects the limits of intuitive exactness and does not allow it to be increased a thousandfold. Still, this hypothesis can be checked in a certain way by ascertaining that the duration of m successive oscillations is to that of n oscillations as m is to n (m and n being large integers)—all within the limits ± 0.1 sec of the exactness of observation, of course. In general the matter is as follows: through the exact laws of the theory which is taken as a basis, the quantity x to be determined is placed in functional dependence on a score of other quantities. By observing the latter, conclusions can be drawn as to the value of x which permits us to determine x more precisely than is possible by direct observation (if this is possible at all). The underlying theory makes good, if within the limit of error all indirect methods of determining x lead to the same result.

Hence, a "right" theory of the course of the world has to fulfill on the basis of our considerations the following demands: 1. *Concordance.* If x is a quantity that occurs in our theory, the definite value to be assigned to x in an individually determined case is ascertained by means of the theoretically established connections and the contact between symbolical theory and immediate experience. All such ascertainments must lead to the same result. Thus all determinations of the charge e of the electron lead to the same value of e (within the limits of observation) if one combines observation with the laws established by our physical theories. One frequently compares one (relatively) direct observation of the quantity under consideration (position of a comet among the stars at a certain moment, for instance) with a calculation on the basis of different observations (present position of the comet, for instance, as calculated by means of Newton's theory from its position during several successive days a month ago). The demand of concordance involves the *consistency* of the theory, but it reaches beyond it since it brings the theory into contact with experience. 2. It must always be possible in principle to ascertain the value of a quantity x that occurs in our theory in an individually determined case on the basis of what is given in experience. In other words: a theory must not contain any parts that are superfluous for the explanation of the observed phenomena.

Since scientific cognition is not a description which faithfully portrays what is given but is rather a theoretical construction, its individual propositions do not possess a meaning verifiable intuitively, but their truth refers to a connected system that can be confronted with experience only as a whole. In Science the word "truth" takes on a meaning which is rather problematic epistemologically and quite different from its application to judgments that do nothing but simply state a fact as it is intuitively given—such judgments as the sentence: this blackboard as it is given to me by perception (beware! I do not talk of the really existing blackboard) has this black color displayed in the same observation. I should like to use the theory of electromagnetic phenomena to serve as an example; but I will simplify matters a bit—as can be done without any essential damage—by putting the speed of propagation of electromagnetic disturbances equal to infinity (whereas its true value is the velocity of light). We assume the existence of particles, "electrons," elementary quanta of matter, endowed with invariable masses and charges. The positions and velocities of these electrons at a moment t uniquely determine the electromagnetic field according to certain laws. Further laws connect this field with the momentum and energy distributed in space; and by means of the flux of momentum the field exercises certain ponderomotoric forces upon the generating particles. Finally the force produces acceleration of the electrons according to the fundamental law of mechanics; velocity and acceleration are the rates of variation of position and velocity respectively during the next infinitesimal time interval dt and hence position and velocity at the moment $t + dt$ are finally determined by their values at the time t. One gets the whole motion by repeating this differential transition $t \rightarrow t + dt$ again and again by means of an integrating process. Only this whole connected theory into the texture of which geometry also is essentially interwoven—is capable of being checked by experiment—provided we assume for the sake of simplicity that the motion of electrons is directly observable (even this is still far enough from the truth). An individual law isolated from this theoretical structure simply dangles in the air. Thus all parts of physics and geometry finally coalesce into one indissoluble unity.

For this reason it happens that broadened or more precise experiences and new discoveries do not overthrow old theories but simply correct them. One looks for the least possible change in the historically developed theory that may

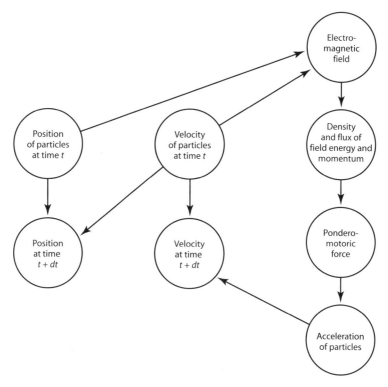

Figure 5.6   How the laws of nature determine the motion of elementary particles.

account for the new facts. Kepler's and Newton's theory of the planetary orbits was founded on the facts observed by means of the tacit assumption that all occurrences are simultaneous with their observation. Later on Roemer discovered, however, the finite velocity of the propagation of light. In fact, he derived it from the apparent deviation of the moons of Jupiter from their theoretically predicted orbits. To construct a theory (the theory of planetary orbits) one uses here the instantaneous propagation of light as a hypothesis. Later more precise observations do not agree; in order to stick to the theory on account of its convincing simplicity the hypothesis must be altered. Rough truth of a hypothesis thus leads to its fine inaccuracy and the necessary correction. But without assuming its rough truth one cannot take even the first step.

If a fact is not in concordance with the entire theoretical stock of science, it is left to the tact and genius of the inquirer to find the weakest point of theory which can most suitably be altered to fit the new facts. Scarcely any general rules can be set up in this respect, as little as for the weight to be given to the several facts (which we know or think to know), for the purpose of their theoretical interpretation. The general relativity theory came into existence when Einstein recognized the fundamental nature and peculiar trustworthiness of the proportionality between weight and inertial mass. One cannot wave aside the possibility of various different constructions being appropriate to explain the observations. In such an acknowledgment of ambiguity of truth Hobbes and d'Alembert precede the modern positivists. In a congratulatory address for Planck in 1918, Einstein characterizes the real epistemological situation very justly as follows:

> Evolution has shown that among all conceivable theoretical constructions there always exists one that proves by far superior to the others. Nobody who really goes into the matter, will deny that the world of perceptions determines the theoretical system practically uniquely— although no logical way leads to the principles of the theory.

In this regard the regulative maxim of simplicity plays a decisive part. But let us abstain from entering into this new important topic.

Instead I want to turn your attention toward a different question. Another feature of physical concepts besides their indirect character is worthy of consideration: they are constructions within a free realm of possibilities. If I am not mistaken, one speaks of a *quantitative* analysis of nature, just for this reason. The impossibility of designing a picture of reality other than on the background of possibility appears to be founded on the circumstance that existence is a penetration of the what and the how, and consequently arises from a contact of object and subject, of pure factuality and freedom. Thus the four-dimensional continuum of space and time is the field of the a priori possible coincidences of events. Indeed, space and time are nothing in themselves, but only a certain order of the reality existing and happening in them. One space point considered by itself is not different from any other. Hence the ascertainment that a given body is found in this position has no contents that could be objectively realized without direct specification "this position here." The two propositions "Body A has this position α" and "Body B has this position α too" contain as their objective part only the one proposition: "Body A coincides with body B." Einstein emphasizes the fact that all physical measurements ascertain coincidences as, for instance, that this pointer coincides with this line on a scale. We cannot restrict ourselves, however, to fixing the individual actually occurring coincidences, but we need rather a field of possible coincidences open to our free mathematical construction. I shall return to this subject when I deal with relativity theory. For this reason Leibniz calls space "the order of all positions which are assumed to be possible," and he adds, "consequently it is something ideal," and in this manner he first recognized the subjectivity of our intuition of space.

The laws of nature can be described only by opposing what is once and for all objectively given to something freely variable, and then seizing upon such elements, the "invariants," which are not affected by this variation. The so-called law of constancy of the velocity of light will serve as an example. The name is not a lucky one. An individual space-time-position or world-point is a here-now, which may be marked by a flash starting up and immediately dying again. They form, as we have said before, a four-dimensional continuum. We consider the manifold of all world points which are reached by a light signal sent out from a definite world point O; this manifold may be called the light cone issuing from O. Now the law of constancy of light velocity asserts nothing else but that this light cone is independent of the state, and in particular of the motion of the light source that sends out the light signal in passing O. This light cone is, however, according to general relativity theory dependent on the distribution of matter within the world and its physical condition. But the light source as something freely variable is here to be opposed to the fixed objective material content of the world.

We try to indicate the distance of the sun from the earth in feet. Such a statement would acquire a meaning verifiable in the given state of facts only if we had a rigid pole extending from the earth to the sun and bearing the individual divisions that had been marked by laying off a movable rigid measuring rod. But this rigid pole does not actually exist, the measurement by a rigid rod is not really carried out; we imagine only that this could be done. Hence geometric statements are ideal determinations referring to mere possibilities of measuring; when taken individually they have no meaning that can be exhibited in actual occurrences. The network of ideal determinations touches experiential reality only here and there, and at these points things must check. Analysis, dissection into "elements," must at any rate be driven so far, as to fix every element in its full concretion by means of the individual value of a quantity varying within a range of possibilities which can be surveyed, since it results from free construction. A plane light wave, for instance, is completely described by its direction, frequency, and intensity, each of which varies within a continuum of possibilities that we can adequately handle by means of the mathematical concept of number. On the whole the series of integers and the continuum of real numbers are the most eminent examples of such an infinite field of free constructive possibilities.

In physics we do not a posteriori describe what actually occurs in analogy to the classification of the plants that actually exist on earth, but instead we apply an a priori construction of the possible, into which the actual is embedded on the basis of the values of attributes indirectly determined by reactions. "Through Copernicus', Kepler's, and Galileo's great discoveries," says Dilthey, "and through the accompanying theory of constructing nature by means of mathematical elements given a priori was thus founded the sovereign consciousness of the autonomy of the human intellect and of its power over nature; a doctrine which became the prevailing conviction of the most advanced minds." But construction a priori must be joined with experience and analysis of experience by experiments. Or, if we once more quote Dilthey:

> The scientific imagination of Man became tamed by strict methods which subjected the possibilities afforded by mathematical thinking, to experience, experiment and verification by facts. The results thus found made possible a continuous and regular progress of scientific knowledge through united efforts of the different countries. One may say, that only since then has human reason become effective as a unified force within the collaborating civilized nations. The most difficult work on this planet was accomplished by thus regulating scientific phantasy and subjecting it to experience.

The simplest and perhaps the most instructive example of transition from description to construction is the creation of the sequence of natural numbers 1, 2, 3, ... At the same time this example is typical for the introduction of symbols. I may hear two sequences of sounds, one after the other. In reproducing the sounds of the first melody by recollection when listening to the second I may ascertain that the second sequence projects beyond the first: "This time there were more sounds than the first time." This statement can be understood without any reference to symbols. However, I may proceed in a different way. While I listen to the sounds I put strokes on paper one after the other, one stroke for each sound. I may thus get the number symbol //// called 4 (four) for the first sequence, and ////// called 6 (six) for the second one, and now I ascertain from the symbols: $6 > 4$. This evidence I do not take from looking at them (a method applicable only to the lowest numbers) but by a certain manipulation: I cross out the first stroke from each symbol and repeat this operation until one symbol is exhausted. Then I find that this is the case with the first symbol 4. One readily recognizes the analogy of this procedure to localization: Instead of simply ascertaining the coincidence of two events A and B we refer to the ideal substratum of space-time-points which can be given only symbolically by means of coordinates: A happens at the space-time-point with the coordinates (t_1, x_1, y_1, z_1), whereas B at the space-time-point (t_2, x_2, y_2, z_2); and then we ascertain the coincidence with reference to the symbols by means of the equations

$$t_1 = t_2; \quad x_1 = x_2; \quad y_1 = y_2; \quad z_1 = z_2.$$

In both cases we are not satisfied to associate with the actually occurring sequences or events respectively, their numerical symbols in the one case, their space-time-points in the other case, but instead we embed the actually occurring number symbols into the sequence of all possible numbers.

This sequence originates by means of a generating process in accordance with the principle that from a given number a new one, the following one, can always be generated by adding the unit. Here the being is projected on to the background of the possible, or more precisely on to an ordered manifold of possibilities producible according to a fixed procedure and open toward infinity. Only then does arithmetic proper come into existence with its characteristic principle of the so-called complete induction, the conclusion from n to $n + 1$. Matters stand similarly with regard to the continuum, the different points of which can be caught more and more precisely by means of a process of division with indefinitely increasing fineness. As Aristotle already observes: "Within the continuum there are, indeed, indefinitely many halves, however not in reality but in possibility only." Mathematics, after all, is not that petrified and petrifying scheme which the layman generally considers it to be; but it stands at that cross point of restrictedness and freedom that is the essence of man himself.

I hope you will understand, if I now describe the essential features of constructive cognition as follows:

1. Upon that which is given, certain reactions are performed by which the given is in general brought together with other elements capable of being varied arbitrarily. If the results to be read from these reactions are found to be independent of the variable auxiliary elements they are then introduced as attributes inherent in the things themselves (even if we do not actually perform

those reactions on which their meaning rests, but only believe in the possibility of their being performed).

2. By the introduction of symbols, the judgments are split up; and a part of the manipulations is made independent of the given and its duration by being shifted on to the representing symbols which are time resisting and simultaneously serve the purpose of preservation and communication. Thereby the unrestricted handling of notions arises in counterpoint to their application, ideas in a relatively independent manner confront reality.

3. Symbols are not produced simply "according to demand" wherever they correspond to actual occurrences, but they are embedded into an ordered manifold of possibilities created by free construction and open towards infinity. Only in this way may we contrive to predict the future, for the future obviously is not given actually.

The problem which a theory of scientific cognition must answer may be crudely stated in this fashion. A comet will find its position of tomorrow by starting out from its present position and by really performing its motion. *We* find its position of tomorrow by drawing certain figures that symbolize the data now at our disposal, by performing complicated symbolical operations on them and thus predicting its future position without any need to wait or the actual performance of its motion. What is it that this symbolical process of the astronomer has in common with the real process of the comet? I do not know whether the considerations propounded to you in my two preceding lectures have been of much help in solving this problem. But this may frequently enough occur to one engaged in philosophical research: as long as he is getting on with his investigation the situation seems to become clearer and better understandable. But when one stops and looks back on the initial problem in its entire primitivity and darkness, one may perhaps feel that it has remained as obscure and puzzling as before, in spite of all the pains and skill employed to solve it. Yet I venture to hope that at least this much may have become intelligible: how and to what extent the structure of our scientific knowledge is conditioned by the circumstance, that the world, the goal of all our scientific endeavors, is not one existing in itself, but arises from and exists only by means of the meeting of subject and object.

May I be allowed to add another observation: concerning the part of logic in scientific cognition. The first science the Greeks set up in a mathematical manner was geometry. After stating the fundamental facts by means of axioms, the further procedure consisted of drawing logical conclusions from those premises; a renewed visualization of the subject of investigation, a presentation of the intuitive meaning of those geometrical objects and relations was not demanded. This one calls the deductive development of geometry. By this procedure geometry has become the model of all strict sciences, and there are not few who consider this to be the main purpose of our teaching of geometry in school: education in strict logical thinking. Today we are aware, however, that this reduction of geometry to logic accomplished only the first step. In building up geometry logically and following its demonstrations one is not presupposed to realize the intuitive meaning of the geometric terms involved, but one must understand all logical terms, expressions like "and," "or," "implies," "all," "exist," and so on. Yet it is possible to free oneself even from this by axiomatizing logic also in a second step. What remains is an operational manipulation of symbols according to definite rules, the symbols representing partly geometrical, partly logical notions. Hence logical thinking and logical inferring is not the core of theoretical procedure as performed in mathematics and the sciences, but rather the practical management of symbols in accordance with certain rules. Of course we scientists have our conjectures and leading ideas; but the strict systematical performance of our method consists of a shoving around of men in a chess game—a chess game, indeed, that proves to be rather significant for reality.

This remark—that lags behind a little because I could not dispose of it elsewhere—may conclude the general methodological consideration in which we were engaged during three lectures; they profess to form a whole the several parts of which are connected and develop organically one from the other. The rest of my time shall be dedicated to illustrating these general observations by means of two of the main doctrines of modern physics: relativity theory and quantum theory. As I scarcely am able to condense what I intend to say about relativity into one hour I beg your permission to start the subject with some historical and systematical preliminaries today.

When one is placed before the task of describing a position P on a plane in a conceptual way (not by means of a demonstrative this-here) one realizes that this can be done only relatively to a system of coordinates—or, if you prefer, to two fixed points A, B, namely by giving the distances AP, BP in terms of a unit of length chosen once and for all. Every point in itself is equal to every other point, there is no objectively tangible property that holds for one and does not hold for the other. In the same sense all directions at a given point are equivalent, and a definite length can be conceptually characterized only by reference to a fixed unit of length. Such are the typical facts with which relativity theory is concerned. The distinction between conceptual fixation and an individual demonstrative act (this-here) is here obviously decisive. Model and source of every demonstrative act is the little word "I." Thus the problem of relativity reveals a new specific side of the subject-object relation. If to every element P there corresponds in a certain realm of objects an element P' of the same realm such that the transition from P to P' destroys no objective relations prevailing among the objects P, we are concerned with an isomorphic representation of a realm upon itself, an "automorphism," as it is called in mathematics. (We must also assume here, that conversely P is uniquely determined by its image P'.) In geometry, the similitudes are obviously such automorphic representations. Figures arising from each other by means of an automorphic representation differ in no respect when judged each by itself; they have all objective properties in common—in spite of being individually different. The group of all these automorphisms expresses the kind of relativity peculiar to a domain of objects in the most appropriate mathematical way.

Relativity of position involves relativity of motion. What we are accustomed to call rest and motion in everyday life, is in most cases rest and motion relatively to the "fixed, well-founded earth." In this sense houses stand still and cars move about. Aristotle already designated position (*topos*) as a relation of one body to the bodies of its surroundings. Locke deals with the matter impressively enough. I quote a nice example from the second book, chapter 13, of his treatise *On Human Understanding*:

> Thus a company of chessmen standing on the same squares of the chessboard, where we left them, we say, they are all in the same place, or unmoved; though, perhaps, the chessboard has been in the meantime carried out of one room into another, because we compared them only to the parts of the chessboard, which keep the same distance one with another. The chessboard, we also say, is in the same place it was, if it remain in the same part of the cabin, though, perhaps, the ship which it is in, sails all the while; and the ship is said to be in the same place, supposing it kept the same distance with the parts of the neighboring land, though, perhaps, the earth has turned round; and so both, chessmen and board, and ship, have everyone changed

place, in respect to remoter bodies, which kept the same distance one with another.

Galileo in his dialogue *Delli due massimi sistemi del mondo*, illustrates relativity of motion quite prettily by a person writing his notes on board a ship sailing from Venice to Alexandretta; his writing pen "in reality," i.e., relatively to the earth, draws a long slightly waving smooth line from Venice to Alexandretta.

In contrast to this Newton at the beginning of his *Principia* proclaims with forceful words the ideas of absolute space, absolute time and absolute motion. But he too, of course, is aware of the fact that one can derive from the observed change of the mutual positions of bodies their relative motion only. His scientific program consists of inferring the true motions of bodies from their relative motions, i.e., the differences of the true ones, and from the forces that cause the motions. In this last respect he depends upon dynamics rather than upon kinematics. Again I quote the author's own words:

> It is indeed a matter of great difficulty to discover, and effectually to distinguish, the true motions of particular bodies from the apparent; because the parts of that immovable space, in which those motions are performed, do by no means come under the observation of our senses. Yet the thing is not altogether desperate; for we have some arguments to guide us, partly from the apparent motions, which are the differences of the true motions; partly from the forces, which are the causes and effects of the true motions. For instance, if two globes, kept at a given distance one from the other by means of a cord that connects them, were revolved about their common centre of gravity, we might, from the tension of the cord, discover the endeavor of the globes to recede from the axis of their motion, and from thence we might compute the quantity of their circular motions. . . . But how we are to collect the true motions from their causes, effects, and apparent differences; and, vice versa, how from the motions, either true or apparent. we may come to the knowledge of their causes and effects, shall be explained more at large in the following tract. For to this end it was that I composed it.[16]

Incidentally, Newton only partly contrives to solve his question: what he is capable of distinguishing from other conditions of motion is uniform translation, movement along a straight line with constant velocity, the pure inertial motion of a body influenced by no external forces; he does not succeed in isolating rest among these translations.

And he must fail on account of the validity of the so-called special relativity principle that is satisfied by the laws of Newton's mechanics and is confirmed today for all natural phenomena by quite a number of the most subtle experiments; all processes going on in the cabin of a boat that sails in a straight line with even velocity, occur absolutely in the same manner as with the boat at anchor. Together with a given process there is always likewise possible that process which arises from it by imparting a common uniform translation to all bodies concerned. This principle was already developed by Galileo in his *Dialogue* in a sufficiently clear and intuitive fashion. Hence one wonders why Newton kept to his conviction of absolute space. This was an empirically unsupported and theologically impregnated a priori belief with him, as is witnessed by many passages of his writings. Space to him is the divine omnipresence of God in nature. In his *Opticks*, for instance, we read that God sees through the innermost of all things, infinite space being, so to speak, his sensual organ, and that he thus conceives them in immediate presence.

About the question of the relativity of motion a violent fight was kindled between Leibniz supporting relativity, and Newton, as whose spokesman Clarke, characteristically enough a theologian, served on this occasion. I intend to open the next lecture by reading to you some characteristic quotations from the letters exchanged between Leibniz and Clarke.

IV. Relativity

Leibniz bases his conviction of relativity of place and motion upon the principle of sufficient reason, which is so characteristic of his philosophical system. In his second letter to Clarke he formulates and explains it as follows:

> The basis of mathematics is the principle of contradiction. An additional principle is necessary with the transition from mathematics to physics, as I pointed out in my *Théodicée*, namely the principle of sufficient reason. It requires that nothing can occur without a reason why it should occur just so, rather than in some other way.

Thus Archimedes had to use a special case of the general principle of sufficient reason when he changed from mathematics to physics in his book on equilibrium. He takes for granted: A lever will be at rest when both sides are equally disposed and when equal weights are brought to the ends of the two lever arms. In fact, there is no reason under such circumstances why either side should sink down in preference to the other.

To this Clarke replied:

To be sure, nothing exists without a sufficient reason why it should be so and should not be otherwise. Hence there is no effect without a cause. This sufficient reason, however, is often nothing else than simply the will of God. When we ask why this particular system of matter should have been created at this particular point of space and not somewhere else, the only reason to be given is the mere will of God, considering that all points of space are quite uniform with respect to matter. If God were never allowed to act without a specific cause (such as the specific cause of the excess weight of a scale), then all free choices would be eliminated and fatalism would be the consequence.

Leibniz strikes back:

There are many ways of refuting the imagination of those who take space to be a substance, or at least something absolute, but I will confine myself to one proof only. I say, then, that if space were an absolute being, there would occur things for which it is impossible to give a sufficient reason—in contradiction to our axiom. I prove it in this way. Space is entirely uniform; without things occupying it, there is nothing in which one point of space differs from another. Now from this it follows that, assuming space to be something in itself other than an order of bodies among themselves, it is impossible there should be a reason why God, keeping the same situation of bodies between themselves, should have placed them in space here and not elsewhere, why, for instance, the whole should not be in reverse and that which is now East be West and what is West, East. If, however, space is nothing else but the order or relation of things among themselves, and is nothing at all without bodies except the possibility of giving order to them, then the two supposed states, the one which actually is and the supposititious transposition, would have no difference at all in themselves. The difference is, then, only to be found in the chimerical supposition that there is a reality of space in itself. Apart from this the two supposed different positions would be exactly the same, two absolute indiscernibles, consequently there would be no meaning in asking the reason for preferring one to the other.

It is precisely the same in regard to time. Suppose someone should ask why God did not create the world a year sooner, and should then go on to infer from the fact that he did not, that God had done that of which it is impossible there could be a reason why he had done it thus and not otherwise. We should have to admit that his inference would be true if time were something outside the temporal things. For it would be impossible that there could be reasons why things should have been set going at such instants rather than at others, their succession when set going remaining the same. What it really proves, however, is that instants apart from things are nothing, instants consist only in the successive order of things. If the successive order remained the same, the two states, the imagined anticipation and the state which now is, would differ in nothing and there would be no way of discerning the difference.

Leibniz illustrates the role space plays for the localization of bodies quite adroitly by means of the example of a family tree. It serves the purpose of expressing the mutual relations of kinship between persons by attributing them a definite position on the branches of the tree. But the tree does not exist before and independently of the men enrolled on the tree.

One clearly observes how this whole controversy is impregnated with theology. Newton and Clarke need God just for that purpose, that he decrees in an arbitrary manner and without inner reasons matters that could not be settled otherwise—whereas Leibniz's idea of God's dignity does not permit him to burden God with such decisions. In this quarrel modern Physics sides entirely with Leibniz.

Physics, therefore, is bound to take all conditions of motion of a body as equivalent. But Newton found that there exist at least dynamical, if not kinematical differences among them: uniform translation is set apart as the movement of a body on which no forces act from outside. This antinomy of kinematics and dynamics demands explanation. There was much fighting over

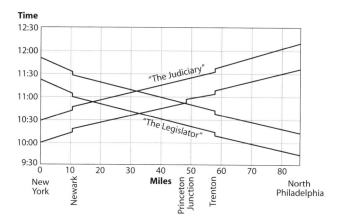

Figure 5.7   Four trains on the track: New York–Philadelphia.

this question during two centuries. The answer as offered by relativity theory cannot be made clear without a little mathematics.

The motions of bodies which move on a horizontal plane E can be represented by means of a graphical picture in the following way. We plot the time t on an axis perpendicular to E in our diagram. An event occurring at the point P of our plane E at time t is depicted, with regard to its position in space and time, by a point in our diagram that is situated perpendicularly over P at height t. This procedure is applied, for instance, in the railroad service where one constructs graphical time-tables of the trains (fig. 5.7). Every small moving body describes a "world line" on which all space-time-points this body passes by are situated. The vertical projection on the horizontal plane E gives the spatial orbit. But one may read from the world line, in addition to that orbit, the temporal law according to which the body moves along. The steeper the line the slower the body moves. The world line of a body resting on the plate E is a vertical straight line. If it performs a uniform translation (relatively to E), i.e., if it moves along a straight line in space with constant velocity, its world line appears as a straight line in our diagram. The meeting of two bodies is indicated by the fact that the corresponding world lines intersect; the point of intersection gives the where and when of that meeting. All events on a vertical straight line coincide with respect to their spatial location, all events on a horizontal plane occur at the same time. The graph of a further occurrence may be drawn: the graph of the propagation of light. A light signal is sent out at the point O on our plane at time $t = 0$. After a lapse of 1 sec it is received at all points, or stations, P at the distance $c = 300,000$ km from O. They form

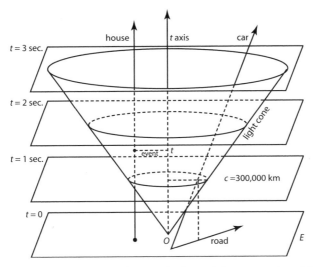

Figure 5.8   Graphic representation of moving bodies and the propagation of light on a plane E.

a circle of radius c around the center O. The measuring units may be so chosen that the same length indicating 1 sec in vertical direction equals this light ray c in horizontal direction. After 2 sec the light will arrive on a concentric circle of double radius and so forth. You have to draw these circles one above the other in the layers $t = 1, t = 2, \ldots$ respectively. Hence the world points where the light signal sent out from the world point $(O, t = 0)$ is received form a vertical circular cone with the vertex angle of 90° in our diagram. I mentioned this light cone before—the locus of all world points where a light signal is observed. The plane is here used as the reference body in terms of which all motions are described—in the same manner as the earth serves us for that purpose almost always in our everyday life. We robbed space of one dimension and we restricted ourselves to occurrences in a two-dimensional plane only for making a graphic representation possible.

The statement that light propagates itself with velocity c in concentric circles around its origin O relatively to E holds only if the condition of movement of E is an appropriate one. He who believes in the substantial light ether would say: E must be at rest with respect to the light ether. In the same way the assertion that bodies when not being influenced by any external force move relatively to E in a straight line with equal velocity, involves a certain requirement concerning the state of motion of our reference plane E. If both requirements be fulfilled we may call E, with Einstein, an allowable body of reference. In order to designate the position on E as well as the time t by means of numbers we have to scratch a rectangular cross of axes into the plane E and to agree upon an individual unit of length. The two coordinates x, y in the coordinate system thus provided, determine a position on E. Every world point is now characterized by three coordinates $(t, x, y) = (x_0, x_1, x_2)$. Every structure consisting of world points can be described arithmetically according to the principles of analytic geometry. Thus the light cone issuing from the world point t^0, x^0, y^0 consists of all and only such world points t, x, y as satisfy the relations

$$(t - t^0)^2 - (x - x^0)^2 - (y - y^0)^2 = 0, \ t - t^0 \geq 0.$$

We are used to measuring space- and time-coordinates by means of rigid rods and clocks. In doing so we use certain physical processes, and we rely upon certain assumptions concerning their laws—processes and laws that divulge an intrinsic structure, the metrical structure of space and time. We are on a safer side when we first refrain from all particular physical hypotheses. This compels us to view the concept of coordinates in an essentially more fundamental manner. Coordinates are not measured any longer; they mean nothing more than an arbitrary numeration of world points; they are just symbols serving the purpose of labeling and distinguishing the world points from each other. Coordinates are merely marks or names of the world points. Each coordinate is a quantity which has a definite numerical value at every space-time point. Since the world is a continuum we shall naturally assume that this numerical value varies continuously with the point. The coordinate is, in other words, a continuous function of position within the continuum of space-time points. This continuum is exhibited as a four-dimensional one by the fact that we need four such coordinates or functions of position x_i ($i = 0, 1, 2, 3$) in order to distinguish a single point of the manifold from all others by means of the values of its coordinates. A four-dimensional continuum when thus referred to coordinates x_i is mapped upon the so-called four-dimensional number space, i.e., the continuum of all possible quadruples (x_0, x_1, x_2, x_3) of numbers. I shall not hesitate to replace this number space by ordinary intuitive space in my following descriptions. But I do this only because it enables me to use a more familiar language that is in compliance with my listeners' customs of thinking and intuition. I could not do this, of course, without discarding one of the four dimensions in mind. But let it be understood once for all, that all geometric terms actually aim at the number space.

As long as I am not concerned with the real things within the world and with their laws, there is no reason why I should prefer one coordinate system to any other. The four-dimensional world apart from its content is merely an amorphous continuum without any structure; only coincidence and immediate neighborhood of space-time positions have immediate significance that can be realized by adequate intuition. Think of the continuum as a mass of plasticine! Only such relations have an objective significance as are preserved under arbitrary deformations of the plasticine. The intersection of two world lines is, for instance, of this kind. The maps of the world traced out by two coordinate systems in number space are related by such a transformation or deformation.

But we now come to the real occurrences and their laws; they disclose a certain structure of the space-time continuum. Newton contends, at the beginning of the *Principia*, that this structure consists of a stratification traversed by fibers. All simultaneous world points form a three-dimensional stratum or layer, the present space, all equipositional world points a one-dimensional fiber. This is the meaning proper of his doctrine of absolute space and time. If such conditions prevail, particular "allowable" coordinate systems may be introduced which are adapted to that structure in a certain manner. In the present case, for instance, one demands that the coordinate x_0, called time, have the same value at all points of a layer, while the other three coordinates x_1, x_2, x_3 keep constant on a fiber.

What is the empirical right of Newton's assumption? By which real occurrences, so we must ask, does he determine that stratification and fibration? As we have seen, his scientific program was, indeed, to answer this question; but he succeeded only to a certain extent. One can distinguish dynamically among all world lines the geodesic ones, i.e., the world lines of bodies that are not subjected to any external forces. A geodesic or a free orbit is uniquely determined by its initial point and initial direction in the world. We call the structure of the universe, to which this dynamical distinction is due, the inertial structure. But one does not succeed as Newton tried in vain, to separate objectively a smaller class within the ensemble of free orbits: the world lines of bodies at rest, among which a single individuum would be determined by the initial point alone (without an initial direction).

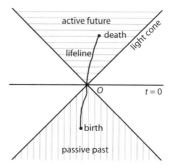

Figure 5.9 Past and future (as determined by the light cone).

What about the layers of simultaneity? Does the belief in their objective significance rest on better foundations? We trust simultaneity since everybody considers without any hesitation the events he observes as happening at the moment of their observation. It is thus that I extend my time over the whole world which comes within my sight. This naive opinion, however, lost its ground long ago when Roemer discovered the finite velocity of propagation of light. So our doubts are stirred up; let us consider the question more carefully. The layer of presence running through a world point O (here-now), is meant to separate past and future from each other. Past and future, what is the reality behind these mysterious words? By shooting bullets from O in all possible directions and with all possible velocities I can only hit those world points which are later than O; I can't shoot into the past, I can no longer kill Caesar. Likewise, an event happening at O has influence only upon the events at later world points; the past cannot be changed. That is to say, the stratification has a causal meaning, it describes, as Leibniz already recognized, the causal connection of the world.

But the modern development of physics leads to corrections with regard to the causal structure which are of disastrous consequences for the old idea of simultaneity. They are due to the experimental discovery that no effect propagates more quickly than with the velocity of light. Hence the above-mentioned light cone with O as its vertex separates past and future in the four-dimensional world at O instead of the plane $x_0 = $ constant. This means that the causal structure has a somewhat different character from what Newton supposed to be true. Let me describe the situation a little more in detail. If I am at O, my life line, the world line of my body, is divided by O into two parts: past and future. This remains true as before. Past and future, as I know them from the experience of my inner life, stay quite unchanged. A different concern,

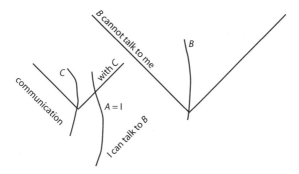

Figure 5.10   Are *A*, *B*, *C* contemporaries? (Light cones issuing from points of my life line, *A*, meet the life line of *B*, but no light cone issuing from *B* meets *A*. *A* sends a light message to *C*, and *C* is able to answer: mutual communication.)

however, is my relation to the external world. World points are located within the light cone issuing from *O*, if and only if they are influenced by what I do or do not do at *O*. Outside of the cone are located all events which lie behind me and which cannot now be altered any more. The cone comprehends my active future. I may complete this forward cone by its backward prolongation to form a double cone. All events that may be of influence upon me at *O* are localized in the backward cone, i.e., all world points *P* such that the forward light cone issuing from *P* contains *O* as an interior point. In particular the backward cone comprises all events which I have seen and witnessed myself or from which I received any message or written documents, that finally go back to eye-witnesses: the backward cone is the domain of my passive past. Both regions, that of active future and of passive past, touch at *O*, without bordering on each other elsewhere; they are, indeed, separated by an intermediate region which I am not connected with at present, either actively or passively. This makes the difference in comparison to the older concepts: they let active future and passive past border on each other along the whole layer of presence.

It is by no means difficult to adjust oneself to this new concept of the causal structure. Consider the question whether a person here and on Sirius are contemporaries. Instead of this question one has to ask more concretely: Am I able to influence him, for instance, to send him a message, or conversely: Can he send me a message, or can we communicate, i.e., can I send him a message and receive his answer? etc. Each of these questions points to a different situation. (Compare fig. 5.10.)

Inertial structure and causal structure have to take the place of Newton's absolute space and absolute time. The insight into the nature of inertial and causal structure—their being different from what Newton conceived them to be—is the first essential part of relativity theory (to which the word "relativity" is, of course, quite inadequate). In the same way as inertial structure is based on the empirical fact, that the geodesic as described by a free body is uniquely determined by means of its starting point and direction (which can be chosen at random), so is the causal structure based on the law of the constant velocity of light. It asserts, as you may remember, that the light cone issuing from O is uniquely determined by O, independently of the condition of the light source that emits the light signal when passing the world point O.

But now back to relativity theory and coordinates! The problem is how to distinguish a particular coordinate system or a whole class of such from all possible systems in an objective manner, not by individual demonstrative acts (as represented by words like "I," "this," "here," "now"). The only possible way is to declare: Such and such physical processes are expressed in such and such an arithmetical way by means of the coordinates wanted. This is the simple content of the famous postulate of general relativity. The special relativity theory contradicts it by no means. It maintains merely that special coordinate systems exist, that the world can be mapped upon number space in such a way that

1. the geodesics appear as straight lines and
2. all light cones are represented by vertical circular cones with the vertex angle of $90°$.

General relativity theory, however, doubts on good reasons that such distinguished coordinate systems exist. And now we finally come back to that discrepancy between the kinematical and dynamical analysis of motion which stirred up all our discussion.

We discovered the inertial structure of the world as being the cause of the dynamic inequalities which prevail among motions. According to Galileo the actual movement of a body is determined by the struggle of two tendencies: the body's inertia and the forces which try to deviate the body from its inertial path. The strength with which inertia resists the deviating forces is measured by the inertial mass. Galileo's and Newton's physics conceive of the inertial structure as a rigid geometric property of the world—in the same way as ordinary geometry fixes the difference between straight and curved lines geometrically, i.e., once for all and independently of material influences. This is revealed by the fact that the arithmetical expression of all inertial movements in terms of a single appropriate coordinate system is absolutely determined and involves no kind of arbitrariness. Here we recognize why the situation appeared so little satisfactory: a thing which produces so enormous an effect as inertia does when it rends, for instance, the cars of two colliding trains by its combat with the elastic forces which act between the molecules of the trains at touch—such a thing is supposed to be a rigid geometric property of the world, fixed once for all. It acts, but it does not react! This is unbearable. Hence the solution is won as soon as we dare to acknowledge the inertial structure as a real thing, that not only exercises effects on matter but is submitted to such effects itself.

Let me illustrate what I mean by a much older instance. Democritus still imagined space to be endowed with an absolutely distinguished direction, that from top to bottom, and believed that bodies when undisturbed follow this direction in empty space. Now, nobody could deny the existence of such a vertical structure of our space; it belongs to our most commonplace daily experiences. But we learned in the meantime that the vertical, the direction of gravitation, is not a geometric property of space but has a physical cause, that it differs at different points of the earth and varies with the physical conditions prevailing here or there. It is influenced by the distribution of matter. The objection so often raised against the doctrine of the spherical bulk of the earth, during the Middle Ages, that our antipodes hang with their heads down from the earth and would fall headlong into the void, lost its conviction for us. Well, exactly in the same sense as in this example of the vertical structure, we must get used to the idea that the inertial structure of the world is not rigid, but flexible, and changes under material influences. This step was taken by Riemann as early as the middle of the nineteenth century, as far as the metrical structure of space is concerned. Einstein rediscovered this thought independently of Riemann, completing it by an important cognition that rendered it physically fruitful: from the equality of inertial mass and weight—a fact known to all and understood by nobody—he concluded that gravitation is not a force, but a part of inertia; it has to be put on the side of inertia in the dualism of inertia and force. The phenomena of gravitation thus divulge the fact that the field of inertia is changeable and depends on matter. The splitting of the uniform field of inertia into a homogeneous part, which is alike everywhere and accounts for Galileo's law of inertia, and a much weaker deviation called gravitation, which surrounds the individual stars—this dissection is not absolute but relative to a coordinate system and hence differs according to the coordinate system.

In the same manner general relativity delivers the causal structure, as represented by the light cone, from its geometric rigidity and makes it dependent on matter. By the way, Einstein reduces both, causal and inertial structure, to a deeper-lying metric structure of the universe, and he does not hesitate to use the old consecrated name of "ether" for it. But we need not enter into this. But keep this in mind as the second fundamental doctrine of relativity theory: inertial and causal structure are something real, of similar constitution as the electromagnetic field, and as such interact with matter. (The first fundamental proposition of relativity theory is, if you allow me to remind you, that the separation of past and future is performed by the light cone, and that the light cone issuing from O is uniquely determined by O, independently of the conditions of the light emitting source at O.)

This description of the essential contents of the relativity theory was necessary in order to attain clarity as to what it states about the relationship of subject and object in scientific cognition. I should like to bring out three points.

Firstly. If we regard the inertial and causal field as something real, and introduce into the physical laws certain quantities of state describing these two fields, it follows clearly from the manner in which they were introduced, that these quantities of state are not directly observable, but that we can only ascertain them, if we consider it to be possible to send out a light signal at every random world point the arrival of which is observed at all space-time places, or to send off at every world point in every random direction a point-mass free from the influence of any forces, the motion of which we follow. On the other hand, however, these "possible" light sources and measuring bodies may not be included in the objective status of the world which is given once and for all; for they must be allowed to vary, and in addition they would modify the distribution of the inertial and causal field which is dependent thereon. The counterview of the definitely given objective world and the observer varying the conditions of his experiment in the domain of the possible is thus here particularly conspicuous.

Secondly. The immediately experienced is subjective but absolute; no matter how cloudy it may be, in this cloudiness it is something given thus and not otherwise. To the contrary, the objective world which we continually take into account in our practical life and which science tries to crystallize into clarity is necessarily relative; to be represented by some definite thing (numbers or other symbols) only after a system of coordinates has been arbitrarily introduced into the world. We said at an earlier place, that every difference in experience must be founded on a difference of the objective conditions; we can now add: in such a difference of the objective conditions as is invariant with regard to coordinate transformations, a difference which cannot be made to vanish by a mere change of the coordinate system used. This pair of counterpoints, subjective-absolute and objective-relative, appears to me to contain one of the most fundamental epistemological cognitions which one can gather from natural science. Who desires the absolute, must take subjectivity, the ego for which things exist, into the bargain; who is urged towards the objective cannot escape from the problem of relativity!

Thirdly. The objective world merely exists, it does not happen; as a whole it has no history. Only before the eye of the consciousness climbing up in the world line of my body, a section of this world "comes to life" and moves past it as a spatial image engaged in temporal transformation. This splitting up of the world into space and time at some moment for some consciousness expresses itself in the following world-geometric construction. At the world point O we have a light cone K and the direction of the world line b, of the observer. In the close neighborhood of O, b may be considered as a straight line. There is a certain set of parallel planes in the neighborhood of O which cut K in similar ellipses whose centers lie on b. If one takes these as strata of simultaneity and the lines parallel to b as fibers of equiposition, one obtains, in the neighborhood of O the decomposition into space and time relative to the observer. The ellipses of section shall be projected parallel to b on to the one plane E of the set passing through O itself: one obtains in it a set of similarly placed ellipses with center at O. Let us call them the gauge ellipses in the spatial plane of the observer, while we shall say of the plane E itself that it is conjugate to the direction of b with regard to the light cone at O (see figs. 5.11 and 5.12).

It seems good, at this place, to describe the relationship of subject and object in completeness with reference to a typical example. It shall be chosen as simply as possible. It is to evidence that the relationship has to be indicated completely, or else one gets stuck in mere verbal definitions of space and time or similar things, which with equal rights can be attacked by the opponent as inadequate with exactly as incomplete arguments. Take the observation of two or more stars of a constellation. I shall simplify the perceiving consciousness

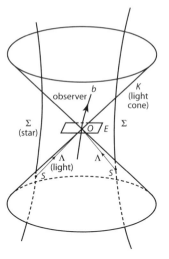

Figure 5.11 Elements on which the angular distance of two stars depends. (World line of the observer b, of two stars Σ, of their light signals Λ. Moment of observation O, spatial neighborhood of observer at this moment E, light cone K.

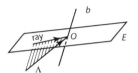
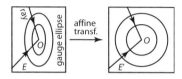

Figure 5.12 Auxiliary constructions. *Left*: Projection of an incoming light signal Λ on the "space" E. *Right*: By affine transformation (= parallel projection) the gauge ellipses are changed into circles.

into a point-eye whose world line shall be called b. The observation takes place at the moment O of its life. The construction must be carried out in the four-dimensional number space of the coordinates; only to make myself more easily understood I shall design a geometric figure. Σ are the world lines of two stars. They intersect the backward light cone K radiating from O in the points S. The light reaching the observer from the stars at the moment O informs him of the condition of the stars at this moment S of their history. In addition, we need the world lines Λ of the light signals which the two stars send from S to the observer at O. These lines lie on the cone K and may be defined as so-called characteristics in the following manner. To every world point P corresponds a light cone $K(P)$, so that we have a field of light cones. A characteristic Λ is a world line which lies on all those cones $K(P)$ of the field that spring from the points P of the line Λ. In the immediate neighborhood of O, the decomposition into space and time with regard to the observer now assumes its rights. We project the line Λ parallel to the direction of b at O onto the spatial plane E through O which is conjugate to the direction b with respect to the light cone $K(O)$. These projections are the spatial light rays. In an auxiliary construction, it is possible by parallel projection to transform the plane E into another plane E' in such a manner, that the gauge ellipses are carried into concentric circles about the center O. By this process, the directions of the two spatial light rays at O are carried into two directions in E'; and it is the angle θ they form with one another which is read from the theodolite as the angle under which the two stars appear to the observer. In these directions for the construction of θ everything is contained: the dependence of the angle on the stars themselves, on the causal field extending between the stars and the observer, on the position of the observer in the world (spatial perspective), and on his condition of motion (different angle values result according to the direction of the line b passing through O: this is the "velocity perspective" known by the name of aberration). These angles θ between any two stars of a constellation determine the constellation's visual form, the form which appears under the assumption that I am the point-eye in question, and which itself cannot be described objectively, in mathematical terms, but can only be experienced in perception. For this reason, one can neither specify in any way the law according to which the mathematical angle magnitudes θ determine the form perceived; the only thing which can be said without reference to the experienced quality is this: If these angles have remained the same during a second observation the constellation will again appear in the same form; if they have changed, in a changed form.

This example, if thought over thoroughly, will lead, I believe, to an understanding of the procedure of physics which we described in general above: It constructs an objective world in mathematical symbols, but afterwards, to establish the connection with experience, it has to indicate by what procedure the quantities are found which are decisive for immediate perception—decisive in such a way, that equal values of these quantities indicate equal perceptions. In doing that it must of course include the observer as a physical being and his condition in the objective world.

In a last lecture, only loosely connected with the preceding, I shall discuss quantum theory with regard to its contribution to the problem of subject and object in physics.

◫ V. Subject and Object in Quantum Physics

Today I should like to throw off the philosopher's gown and tell you a story, the history of the development of quantum theory in its principal stages. I must altogether forego convincing you that the development had to go this way; for this purpose it would be necessary to acquaint you much more completely with the physical facts and to examine much more precisely the theoretical tools and their possibilities. You may believe me that the men who gave the development its decisive impulses were conscientious, and not reckless revolutionaries, that they upheld the beautifully harmonious and complete classical theory of electrodynamic and kinematic phenomena as long and as tenaciously as in any way possible, and that they gradually shifted to another course not out of a mere desire for innovation, but under the irresistible pressure of experiences. Not the seductive game with novel possibilities that one cares to follow for a while until they run into absurdity—as seems to have been the case in modern painting—but bitter necessity has led us to this strange quantum physics. It is difficult to understand because it contradicts certain fundamental conceptions that are firmly anchored in our language, and even today we physicists are not yet quite sure whether we have really understood the situation. What we possess is a mathematical apparatus which functions reliably, which leads to unambiguous predictions everywhere where the experiment is able to decide. Quantum physics in a similar sense represents a crisis of the old idea of causality, as the theory of relativity shook the foundations of space and time. Here too we shall have to discuss the principle of indeterminacy of quantum theory; but we are interested less in its negation of strict causality than in the circumstance that through this principle there opens up between real process and observation, between object and subject, a gap that is essentially deeper than in classical physics.

In the eighteenth century, Huyghens set up the wave theory of light against the corpuscular theory due to Newton. It is the phenomena of interference and diffraction which clearly prove the wave-like nature of light. If a wave of amplitude u meets with a wave of amplitude u', they can mutually reinforce or weaken one another according to the difference in phase existing at the respective place between the two oscillations. The intensity or energy of a wave is proportional to the square u^2 of the amplitude, the intensity produced by the mutual action of the two waves with phase difference δ is computed by the cosine theorem of plane trigonometry to be

$$u^2 + u'^2 - 2uu' \cos \delta.$$

By the combined action of two waves of equal intensity there thus arises at certain places the quadruple intensity of the single one, at certain other places, however, the intensity zero, with all possible intermediate stages. But according to the corpuscular concept by which light consists of ejected particles and the intensity is determined by the energy of these particles, there should arise from the combined action of two fields of radiation of intensity 1 a field of radiation that has everywhere the double intensity. The oscillations generating the light were originally conceived as mechanical oscillations of a substance, of the light ether. This led to more and more serious difficulties and was irreconcilable, as a matter of fact, with the special relativity principle which had been confirmed exactly by experience also in the optical domain. All difficulties vanished and great progress was brought about in the simplification of our physical concepts when Maxwell recognized light as electromagnetic oscillations of high frequency. The Maxwell equations first showed the possible existence of electromagnetic waves, which were then really detected by the experiments of H. Hertz and which find their practical application in wireless telegraphy today. One must have before one's eyes the successes that wave optics gained during the course of almost two centuries in order really to appreciate what weight this theory possesses.

And still, upon penetration into the atomic processes since the beginning of this century, light, to the great surprise of the physicists, again began to reveal also corpuscular traits. At first difficulties arose as one tried to investigate the interaction between matter and radiation with the same means that had been used with such great success since Daniel Bernoulli in gas theory. It was incomprehensible on this basis that at a given temperature any equilibrium at all took place, that at a certain temperature a body, for instance, assumed a state of red-heat, at a higher temperature one of white-heat, with a distribution of energy that was exactly determined in the spectrum of the light and heat rays. The whole energy of the radiation should instead have wandered into the highest frequencies of the emission of which the respective substance is at all capable according to the constitution of its atoms. The contradiction could be solved only by the assumption that the exchange of energy between the atoms and a light wave of frequency ν can take place only in a discrete, not in a continuous manner, namely in integral multiples of an energy quantum ϵ; this is not independent of the frequency v, but is proportional to v for the various frequencies: $\epsilon = hv$. This was the epoch-making discovery of Max Planck in the year 1900. The absolute constant of nature h is exceedingly small:

$$h = 6.547 \times 10^{-27} \text{ erg-sec,}$$

so that one understands very well that for most phenomena the quantic behavior of energy is completely to be neglected. After all the quantum of action h occurred here only in the thermodynamic law of radiation, the law of a process in which the atomistic structure of matter and the quantic structure of radiation is extinguished past recognition by the interaction of innumerable particles. But according to the connection in which it occurred and according to its order of magnitude this h had to be an atomistic constant.

It was therefore a daring step of Einstein in the year 1905 that he seriously applied the concept of the light quantum or photon to the actual atomic processes. The question concerned the so-called photo[electric] effect regarding which investigations by H. Hertz and others were extant. If a metal plate is irradiated with ultra-violet light, the plate ejects electrons. Strangely enough the vehemence with which the electrons are torn out of the plate is not dependent on the intensity of the radiation, but only on the color, that is, the frequency v of the ultra-violet light; it grows with growing v. The intensity is merely of influence upon how many electrons are released per time unit. Here now we have an atomic process which has offered the most obstinate resistance to comprehension by the wave theory of light. The light energy incident on the atom is evidently made use of to tear an electron out of the atomic union and in addition to impart to it its kinetic energy. In the case of weak intensity of the incident light, however, the light wave sweeping the electron is by far not sufficient to produce such an effect; even if one imagines a mechanism in the atom which continually draws energy from the light wave and stores it in the atom, hours would have to go by before the electrons could start to leave the plate. Instead of that the effect takes place immediately. But if we conceive light to consist of individual photons of energy hv, a photon could, by hitting an atom and knocking an electron out of it, place its own energy at the disposal of this electron in the form of kinetic energy. The latter would therefore, in agreement with experience, be dependent only on v and would increase proportionally to v. The intensity of the light merely indicates the density of the photon hail. One measures the current of electrons that is discharged by applying an anti-potential V which checks the electrons. The electron of charge e performs the work eV as it traverses the potential drop V. According to the Einstein theory the current should vanish as soon as the potential exceeds the value $V = \frac{h}{e}\nu$. Not until a decade later had the experimental technique progressed so far in the hands of R. A. Millikan that the decisive measurements could be carried out. It actually becomes manifest that the limiting potential V is proportional to the frequency ν of the incident light and that the constant of proportionality has the value h/e, quotient out of the quantum of action h known from the Planck radiation law and the elementary charge e derived from other investigations. Vice versa, if in a cathode tube with electrons flying from the cathode to the anti-cathode against a potential V, the electrons are checked at the anti-cathode, the tube emits only such X-rays whose frequency lies below the limit obtained from Einstein's equation $\nu = \frac{e}{h}V$; here the continuous X-ray spectrum of the tube has a sharp edge. Thus experience verifies that radiation energy of frequency ν is taken on and cast off only in quanta $h\nu$.

In the meantime Niels Bohr (1913) had founded his theoretical determination of spectra on the same rule. The password for atomic theory seemed to have been found. Millikan begins the report "Recent Developments in Spectroscopy," which he made in the year 1927 before the American Philosophical Society, with the words: "Never in the history of Science has a subject sprung so suddenly from a state of complete obscurity and unintelligibility to a condition of full illumination and predictability as has the field of spectroscopy since the year 1913." According to classical electrodynamics the atom was to radiate continually in consequence of the electronic motions taking place within it; through the radiation itself the atom loses energy, therefore the process of motion is modified and together with it the radiated frequencies are displaced.[17] It is impossible to see from here how the unchanging constant properties of the atoms are to be understood; thus in particular the existence of sharp, time-after-time reproducible spectral lines. The Bohr theory assumes that the electrons can move in certain stationary orbits in which they do not radiate. The atom is then on certain energy levels $E_1, E_2, \ldots$. Light is emitted upon transition from one stationary state into the other; the energy lost in this process, the difference between the two energy levels E_1, E_2 in the two stationary states, is transformed into a light quantum $h\nu$; hence the frequency ν of the emitted spectral line is determined from the equation

$$h\nu = E_1 - E_2.$$

In absorption the opposite takes place.

This rule accounts forthwith for the Ritz-Rydberg combination principle that was gathered from an enormous empirical material; it states that together with the frequencies

$$v(i \to k) = \frac{1}{h}(E_i - E_k) \text{ and } v(k \to l) = \frac{1}{h}(E_k - E_l)$$

the "combined" frequency

$$v(i \to l) = \frac{1}{h}(E_i - E_l) = v(i \to k) + v(k \to l)$$

can also always occur in the spectrum. Bohr's theory also indicates a rule applicable in many cases for the determination of the energy levels; but this was unmistakably a compromise. Nevertheless one succeeded in this manner in interpreting completely the series formula in the hydrogen spectrum and in setting up a connection between the occurring empirical constants and the fundamental atomic constants, charge and mass of the electron, velocity of light and quantum of action. Later more exact measurements have always only been able to confirm this connection more precisely.

Thus light on the one hand exhibited itself in an evident manner as a wave process and on the other hand showed in the atomic happenings that it consists of individual quanta, photons, whose energy content ϵ is bound up with the frequency v of the wave by the relation $\epsilon = hv$. Vice versa, inasmuch as electronic rays produce interferences exactly like X-rays in traversing crystals, Davisson and Germer, in the year 1926, identified an undulatory character in the electrons which after all are doubtless corpuscles. These experiments were already in progress when Heisenberg, L. de Broglie, and Schrödinger arrived at the new conception of quantum mechanics. Heisenberg replaced the compromise by which Bohr had determined the energy levels by a rule which fitted itself perfectly into the Ritz-Rydberg combination principle of frequencies; for this purpose it was, however, necessary to give up completely the intuitive conception of electronic orbits in the atom and to develop a novel mathematical apparatus, the algebra of matrices: De Broglie and Schrödinger again indicated how with every corpuscle there is to be connected a wave capable of interference which directs the behavior of this corpuscle. The actual concordance of both theories which looked so different formerly was soon discovered. As regards the final physical interpretation of the resulting calculatory apparatus M. Born and Dirac—deserve particular mention.

The question was to grasp this double nature of light as well as of the elementary constituents of matter: To be a wave capable of interference and at the same time suddenly to strike here and there as a discontinuous quantum. We meet this double character everywhere in the atomic happenings. The solution can perhaps for the moment be described quite generally as follows: We take the wave theory for a foundation, make use of the quantity of state ψ depending on space and time which it deals with, and the linear differential equations which ψ satisfies and which are in agreement with the principle of causality. If such a ψ is capable of complex values it guarantees the capacity of the waves to superpose with arbitrary phase displacements. But we interpret the theory differently. That quantity ψ^2 which occurs in wave theory as the intensity of the wave at a certain place, and which depends on ψ in a quadratic fashion, shall be considered as the (relative) probability that the particle, photon or electron, is found at this place at a given moment. Or more precisely as the relative probability that it is found in a small volume element about this place which is assumed equally large at all places. The probability of an event can be controlled empirically only by making a large number of experiments and ascertaining the frequency with which the desired event takes place among them. The probability of the birth of a boy is gathered from the relative frequency with which the births of boys are represented among all births. Our conception then is this: ψ represents a certain state of the photon or electron which can be brought about in a manner to be described more precisely. In this state, however, the photon is not necessarily found at a certain place, but if we always produce anew a photon in this state we shall find it once here, once there, without being able to predict anything more exact about it. What we can predict from the state is only the probability, the relative frequency with which we shall find it at this place or the other if we repeat the experiment a very great number of times. Thereby we must not forget that a probability always determines the frequency in a great number of experiments only with a certain factor of uncertainty. It can after all happen once in a sequence of experiments that the frequency departs noticeably from the a priori probability. Thus we do not know and can not know what the individual photon or electron does under given conditions; we can only, with the uncertainty adhering to statistics, predict from the wave image their average behavior under the same conditions. It is a different thing whether we repeat the same experiment time and again with the same photon, or whether we subject a large swarm of photons simultaneously to the experimental conditions; for the photons of the swarm could influence one another. Quantum theory as developed even shows that they by no means behave as though they were statistically independent of one another. Nevertheless it will perhaps simplify the concept and the mode of expression, if we speak of such a swarm of photons instead of an experiment continually repeated on individual photons. The probability then simply appears as the spatial density of the photons.

It was a little daring when we said: We do not know more and cannot know more about the photons than their statistical behavior upon frequent repetition of the experiment under equal conditions. But that is precisely the decisive feature about the new quantum physics that it does not admit of the possibility to complement our theoretical image in such a fashion that we find out more about the individual electron from the laws of nature; it makes it impossible to consider our wave equation as possibly only a part of the complete exact laws which determine the exact behavior of the photon if it is subjected to this or that set of exact conditions. To justify this we shall compare the new quantum physics with the old statistical physics.

The penetration of statistics into physics lies back already more than a century. It took place at first in the kinetic gas theory thought out by Daniel Bernoulli which was developed to high perfection in the nineteenth century by Maxwell and Boltzmann. In 3 gm hydrogen about 10^{24} hydrogen molecules whir around. It is of course in practice impossible to compute exactly on the basis of the mechanical laws of kinematics the motion which they perform under the influences due to the walls of the container and their mutual forces. Also this is not at all what one would like to know in the first place about the behavior of the gas. For observation, only certain mean values are decisive and accessible; so for instance the average kinetic energy of a gas molecule determines the temperature, the bombardment of the molecules against the walls or the impulse thereby transferred in the mean on the surface unit of the wall causes the pressure. And the theory of probabilities permits the computation of such mean values without its being necessary to follow the motion of the innumerable molecules in detail. But besides the mean value of a quantity it also furnishes the mean deviation from this mean value to be expected. Thus it is, for example, the momentary density vacillations of the air occurring in the accidental whirring jumble of the molecules which by deflection of the sunlight make the daytime sky appear blue instead of black.[18] The vacillations, then, taken all together do have an observable effect, however trifling each individual one is. Such vacillation phenomena are above all a support of the statistical conception. It has become clear by the efforts of Boltzmann and Maxwell that none of the notions and laws referring to the thermodynamic behavior of matter are exact ones, but that they are mean values and statistical regularities affected with certain indeterminacies. The general epistemological attitude toward statistical physics was at first absolutely this, that the theory of probability was considered as an abridged way only for arriving at certain consequences of the exact laws of motion; these laws in truth, that was the opinion, regulate the process down into its finest details. Thus, for example, one tried to prove on the basis of classical mechanics that the periods of time during which the gas is in a state deviating perceptibly from thermodynamical equilibrium vanish as compared to the total length of observation, in the limit for an infinitely long period of observation.

In opposition hereto the statistical indeterminacy which adheres to the statements of quantum theory concerning observable processes is an essential one. It is impossible to describe the underlying process in a space-time image exactly in such a manner that our statistical statements are incomplete conclusions from the exact laws. I should like to begin by illustrating this with reference to the polarization of light. All light quanta in a monochromatic rectilinear beam of light have the same exactly determined energy $h\nu$ and the same impulse. If we make the beam pass through a Nicol that has a certain position s (a direction in space orthogonal to the propagation of the light), we impress upon it in addition a certain direction of polarization.[19] In the language of light quanta this will express itself as follows, since a light quantum either will pass through the Nicol or will not: There is attached to the light quantum a quantity p corresponding to the position s of the Nicol which is equal to $+1$ or -1 according as it does or does not pass through. The basic considerations can be exhibited much more emphatically on such quantities of state as are capable of only two values than on those otherwise occurring in physics, like, for example, the position of a proton, which vary within a continuous scale. The polarized, monochromatic plane light wave is the extreme in homogeneity that we can attain. Still we see that such a homogeneous beam of light when sent through a Nicol of the orientation s' different from s, is again split up into a beam that passes through and one that is deflected. The relative intensities of the two partial beams depend in a simple fashion on the angle between the two orientations s' and s; they are the probabilities that for a photon of property $p = 1$, the quantity p' equals $+1$ or -1. One might hope that the light which has passed through the second Nicol consists of photons for which both

$p = 1$ and $p' = 1$. But this contradicts the stated fact that the homogeneity of a monochromatic polarized plane light wave cannot be increased any further. And we actually do find that this light is of exactly the same constitution as light that has only passed through the second and not at all through the first Nicol. Thus the second Nicol destroys the result of the selection $p = 1$ performed by the first Nicol. With reference to the light quantum we can speak significantly of the quantity p, because there is a method for determining the value of this quantity which $= \pm 1$. In the same way we can speak of the quantity p'. But we cannot ask in a significant manner which values the two quantities p, p' have *simultaneously* for a photon. For the measurement of p' or the selection of the photons with $p' = +1$ destroys the possibility of the measurement of p or the selection of the photons with $p = +1$.

This impossibility is not a human deficiency, but is of an essential nature. Another example may perhaps make this clearer. A silver atom has a magnetic momentum, it is a small magnet of a definite force and orientation which we represent by an arrow, the vector m of the magnetic momentum. In a certain space direction z, for example the vertical one, the momentum has a component m_z equal to the orthogonal projection of m on to the direction z. One finds that m_z is capable only of two values that are equal except for sign and which when measured in a certain unit of measure, the magneton, can be put equal ± 1. By applying an inhomogeneous magnetic field acting in the z-direction through which a swarm of silver atoms flies, one succeeds in singling out of the atom beam the two partial beams for which $m_z = +1$ respectively -1. That is a famous experiment; it was performed first by Stern and Gerlach. But what was said here for the z-direction holds for the component in every random direction of space! A vector, however, whose component is ± 1 in every direction is a geometrical nonsense. The solution of quantum theory lies in the following: When the z-component, by selection, has been made to assume a definite one of its two values, the remaining components cannot be determined, only probabilities can be indicated for their possible values ± 1.

We are here confronted with an unexpected limitation of the principle which we developed in the third lecture: That we regard the result of the measurement read from a reaction as a property pertaining to the body under observation in itself, if the result of the measurement does not change upon change of the conditions of the reaction; the assumption being that this body every time enters into it in the same state. Properties like "red," "round," which are immediately given in intuition can significantly be combined by the logical particle "and" to a new complex property "red and round." In quantum theory we recognize that this is not possible for those attributes occurring in physics which exhibit themselves only by interventions and reactions on the basis of laws of nature that are postulated to be valid. For it can happen that the performance of the reaction serving to measure the first attribute makes the measurement of the second attribute impossible in principle. We are here very vigorously admonished not to be too easy going about taking flight out of reality into the realm of "possibilities."

Classical physics demanded that conditions be procured which guarantee the extreme degree of homogeneity; and it assumed that under such conditions every physical quantity concerned had a well-defined value that could be reproduced time after time under the same conditions. Quantum physics too demands that the experimenter produce the "pure state," the homogeneity of which can no more be increased. In the case of light that was the plane monochromatic polarized beam. But the ideal of classical physics is unattainable for it. It cannot ask: Which value does this physical quantity assume in this pure state? but only: With which probability does this physical quantity assume any prescribed value in this pure state? The criterion whether the extreme measure of homogeneity has been reached is the same in classical as well as in quantum physics. The experimental conditions B shall have for effect that certain physical quantities about the entity under examination time and again assume one and the same value under the same conditions B so that the experimental conditions determine the value of these physical quantities. The experimental conditions B' guarantee a higher degree of homogeneity if every physical quantity whose value is determined by B in a reproducible fashion, is also fixed by B' and *nota bene*, at the same value, but if there are physical quantities whose value is fixed by the experimental conditions B' but not by B.

One could devise the following expedient and say: The wave field governed by strict laws is the real thing. How amiss that is is shown right away by the following consideration: If we are dealing with two electrons, we can inquire as to the probability that the one is found at the place with the coordinates (x_1, y_1, z_1) the other at the place with the coordinates (x_2, y_2, z_2). The wave function ψ determining this probability must therefore be a function of the two positions in space $x_1, y_1, z_1; x_2, y_2, z_2$, or the wave in this case does not extend in the usual three-dimensional space but in a six-dimensional space. The more particles are added, the higher the number of dimensions of the space rises in which the de Broglie wave process takes place. This alone should show sufficiently that the wave field is only a theoretical substructure. And—this is the decisive point—the fact remains that this wave field determines the observable quantities only as a priori probabilities determine statistical frequencies; from their linkage the indeterminacy can essentially not be expunged.

I must try to describe this a little more definitely and will choose for that purpose the example of the photons which I quoted already above. Let us take one single photon in a definite condition of polarization o. Its state is represented by a certain plane wave, a solution of Maxwell's electromagnetic equations. According to the principle of superposition valid in the domain of waves one can consider this state as linearly superposed out of two waves polarized in the directions α and $90° + \alpha$ just as one can consider a given vector as the resultant of two perpendicular vectors. It is also possible in reality, by means of an appropriate instrument, the polarizer, to resolve the homogeneous beam of light polarized in the direction o into two beams with the polarization angles α and $90° + \alpha$. What happens then to the single photon? The wave picture says that it is simultaneously in the one and in the other state of polarization, and precisely with the relative strength $\cos^2 \alpha$ and $\sin^2 \alpha$. But observation naturally shows that the photon is either completely in the one or completely in the other beam; for the photon is something indivisible, and observation of course is right. If one repeats the experiment, however, one will find the photon, after its passage through the polarizer, once in the one beam, once in the other, without being able to predict what will happen each time; but, given a very great number, N, of repetitions, it will happen approximately $N \cos^2 \alpha$ times that one meets the photon in the polarization α, and $N \sin^2 \alpha$ times in the polarization $90° + \alpha$. One must renounce giving an intuitive space-time description of what the photon does during the course of the process; but the wave picture of quantum theory is sufficient to predict what is to be expected from the actual observation and measurement.

Or let us now return once more to the example with reference to which I made clear the breakdown of Newton's corpuscular theory of light at the beginning of this lecture—the phenomenon of interference. It will immediately become evident how much the quantum theory of photons differs from it. A monochromatic beam of light shall be decomposed in any way into two components of equal intensity which are later brought to interfere. Following the wave picture quantum theory will say: After the resolution into two components every photon is, so to speak, partly in the one and partly in the other component; every photon interferes only with itself. Therefore a measurement which permits to state the occurrence of the photons in the one or the other beam, a measurement of the energy of the one component, of necessity destroys the capacity of interference of the two components. That is absolutely in agreement with the facts. The relative intensities of the wave field at the various places of space give us the probability with which we may expect the photon here or there. In this sense one can say that nature follows the wave picture as long as one leaves it to itself, without disturbing it by inquisitive observation. But if we first poke in our nose, if we want to know whether the photons are in the one or the other component, we destroy the interference and everything collapses.

We gather from this the following:

1. *The indeterminacy cannot be eliminated.* For after having passed through the polarizer the photon can only be found either in the one or in the other beam. To this discrete alternative is opposed the fact that the states of the photon form a continuous manifold as follows from their capacity of superposition. If there are no intermediate cases, a continuous bridging over of this either-or seems conceivable only in such a manner that there exists a probability p or $1 - p$ respectively for the one and the other case which varies continuously with the state between 0 and 1. In concordance herewith the mathematical scheme of quantum physics shows no gap that would let it appear conceivable that the picture might still once be filled in so as to become one which comprehends the processes strictly causally in their details.

2. *An observation is necessarily connected with an abrupt uncontrollable intervention.* The "objective happening" thereby every time tears off. As long as it was not observed that the photon was present in the one or the other of the two components, its state is described by a wave in which the one and the other component is represented with a certain relative strength. Once the observation has furnished a definite result this wave must of course be replaced by one which represents only this component. The observation itself has no room in the wave picture which renders the physical process in such a manner that it leads to correct predictions concerning the observations. To the observation there corresponds rather the transition from the wave picture to its statistical interpretation, to the probabilities determined by it. In our example I can naturally include in the physical entity under consideration also the instrument of observation and even my eye and the interaction between them and the photons, and design a quantum picture of this whole aggregate and its state.

But this again will merely serve to make predictions about observations which are now carried out on this whole system within which the previous observation has become a non-observed real process. In the measure in which mind, knowledge penetrates in order to make use of the processes as observations for the purpose of interpretation, they lose their lawfulness and controllability. If, on the other hand, we try to untie the real world from the observations, we are left with only a mathematical scheme. Quantum physics necessarily arrives at this decisive insight into the relationship of subject and object. Similar contentions have already frequently been made from philosophical quarters. In contrast hereto physics is remarkable for the definite, mathematically precise form with which this idealistic standpoint finds its expression in the physical theory. One can say that in nature itself, as physics constructs it theoretically, the dualism of object and subject, of law and freedom, is already most distinctly predesigned.

Niels Bohr has recently collected four essays under the title "Atomic theory and description of nature" in which, in a cautious language, comprehensible also to the layman, he probes the situation that has gradually been brought about through quantum theory; if anyone has the calling to do so it is Bohr who has been the leader in this whole development and who at no stage allowed himself to be deceived by the successes attained about the remaining fundamental difficulties. I recommend these lectures emphatically for your reading. I should like to borrow the concluding word of my last lecture from him:

> If a physicist touches such questions [like that concerning the relationship of subject and object] he may perhaps be excused by the circumstance that the new situation given in physics reminds us so insistently of the old truth that we are both spectators and actors in the great drama of existence.

1
Hermann Weyl as a student
at Göttingen, about 1904.

2
Hermann and Hella Weyl in 1913, at the time of their marriage.

3
Hermann Weyl, about 1920.

4
Hella and Hermann Weyl at
the Alhambra, Spain, 1922.

5
This portrait of Hermann Weyl was included in an album presented to David Hilbert in 1922, during the period in which Weyl had turned toward the intuitionistic approach to mathematics advocated by Brouwer. Hilbert had been disturbed by Brouwer's rejection of much of modern mathematics; Weyl's attraction to this approach also troubled Hilbert.

6
Weyl at his desk in Zurich, 1927.

7

David Hilbert and Hermann Weyl, about 1930.

8

Gathering at the Gasthof Vollbrecht in 1932: (*left to right*) E. Witt, Paul Bernays, Hella Weyl, Hermann Weyl, Joachim Weyl, Emil Artin, Emmy Noether, E. Knauf, C. Tsen, E. Bannow.

9
Hermann Weyl on a
seesaw at the Gasthof
Vollbrecht, 1932.

10
Michael and Hermann Weyl, at the time of Michael's
graduation from Princeton, in 1937.

11
Weyl carrying his
briefcase on the beach in
New Jersey, 1937.

12
Hermann Weyl and
André Weil at Princeton,
about 1940.

13
Hella Weyl at Princeton, about 1942.

14
A family gathering in Alexandria, Virginia, about 1942: (*left to right*) in front, Hermann's daughter-in-law Martha (Sonja), his son Joachim, his granddaughter Annemarie; behind them, Hermann and Hella Weyl.

15
Hella Weyl, about 1945.

16
Hermann Weyl and his grandson
Peter at home in Princeton in 1949.

17

Albert Einstein's seventieth-birthday symposium at Princeton in 1949: (*left to right*) H. P. Robertson, Eugene Wigner, Hermann Weyl, Kurt Gödel, I. I. Rabi, Albert Einstein, Rudolf Ladenburg, J. Robert Oppenheimer, G. M. Clemence.

18

Hermann and Ellen Weyl, Zurich, 1950.

19
Hermann Weyl at the beach in Skagen, Denmark, 1951.

20
Hermann Hesse and Hermann Weyl at Sils Maria,
Engadin, Switzerland, about 1953.

21
Hermann Weyl with his grandsons
Peter and Thomas, at the Gornergrat,
near Zermatt, with the Matterhorn in
the background, about 1954.

22
Hermann Weyl, Zurich, 1955.

6 ▣

Address at the Princeton Bicentennial Conference

1946

[1]   During the past three days the various speakers of this conference have discussed the actual state and current problems of our science in all its various branches, and have tried to prolong the lines beyond the present into the future. Will you now lend me your ear for a brief spell in which I shall evoke the past and let my memory roam over some of the outstanding mathematical events of my lifetime? I have reached the age where one likes "to the sessions of sweet silent thought, to summon up remembrance of things past," and is prone to believe that some lesson for the future may be drawn from comparing past with present. A deceptive belief, I hasten to add; what history can do to us is, as Jacob Burckhardt once said, not to make us more clever for the next time, but wiser for all time. I can give a little more life to my review if you will permit me to intersperse it with remarks about the influence of the mathematical events on my own work. The selection will be very subjective anyway. The balance will be weighted in favor of what happened in Central Europe, since I spent most of my productive years in Germany and Switzerland, and also in favor of youth: by the perspective that always has given rise to the myth of the good old times, my younger years seem to me to have been more crowded than the later with important events.

[2]   I wonder whether the organizers of this conference, when they assigned to me the task of talking to you about mathematics in general after the battle is over, had in mind the opening passage of Hardy's charming little book *A Mathematician's Apology*: "It is a melancholy experience for a

professional mathematician to find himself writing *about* mathematics. The function of a mathematician is to do something and not to talk about what he and other mathematicians have done." "I write about mathematics," he continues a little later, "because like any other mathematician who has passed sixty I have no longer the freshness of mind, the energy or the patience to carry on effectively with my proper job." If I view the situation in which I find myself tonight a little less melancholically, it is not because I disagree with Hardy in that "mathematics is a young man's game," but because I do not quite share his scorn "of the man who makes for the man who explains." It seems to me that in mathematics, as in all intellectual endeavors, both things are essential: the deed, the actual construction, on the one side; the reflection on what it means on the other. Creative construction unguarded by reflection is in danger of losing its way, while unbridled reflection is in danger of losing its substance.

[3]   It was my good fortune that I took my first steps in mathematical research under the eyes of a master who combined both sides, mathematical creation and philosophical reflection, to an unusual degree, David Hilbert, who was then at the height of his productive power. So it happened that the first outstanding mathematical event of my life was the development of the *theory of integral equations* by Hilbert. Fredholm's great discovery lay before my time. What could have been more natural than the idea that a finite set of linear equations describing the motion of a discrete set of mass points gives way to a linear integral equation when one passes to the limit of the continuum? And yet science had to travel a long and tortuous road from Daniel Bernoulli's analysis of the vibrating string in 1730, before this general idea was conceived. But slow travel has its compensations: things become more concrete. Ideas ripen only in conjunction with the development of the concrete problems which they are destined to illumine; and that is good so. Fredholm treated the integral limit of linear equations, Hilbert that of a quadratic form. Its transformation onto principal axes led to a general theory of eigenvalues and eigenfunctions.[1] Bernoulli's heuristic procedure of passing from a finite number of points to a continuum was converted into a mathematical proof. But shortly afterwards Erhard Schmidt gave a beautiful direct proof based on the same ideas as Daniel Bernoulli's and Gräffe's method for the computation of the roots of an algebraic equation and Hermann Amandus

Schwarz's construction of the gravest eigentone of a membrane. Hilbert did not yet use the axiomatic formulation in terms of what we now call a Hilbert space, though he exploited to the full the equivalence between the space of square integrable functions and of square summable sequences. By following the growing sequence step by step he established a general theory of bounded operators with their line and band spectra. Soon more direct proofs were found for his results in this wider field too by F. Riesz and others. Their further development, two decades later, under the impact of quantum mechanics, is a familiar story.

[4]   One of the most interesting applications of integral equations made by Hilbert himself is to the solution of Riemann's problem: Given a finite number of singular points in the complex plane, determine n analytic functions that by analytic continuation around each of these points undergo given linear substitutions. Again the simplifier followed on Hilbert's heels, Plemelj, who analyzed the problem as fully as the special case of algebraic functions on a Riemann surface with n sheets had been analyzed long before. Important as the contributions of these simplifiers were, E. Schmidt, F. Riesz, J. Plemelj, I think the true moral of such happenings is this: the discoverer who first breaks through often does rough work, and yet to him belongs the highest honor. Only after the lock is broken, one can study it at leisure and construct a key that opens it more smoothly. From a purely logical standpoint there is no reason why the streamlined methods should not be invented straight away; but in the face of a profound problem man is seldom that clever.

[5]   The first discovery of my own of some consequence is closely connected with Hilbert's theory of eigenvalues. At a Göttingen conference on statistical thermodynamics in which H. A. Lorentz, A. Sommerfeld and others participated, the physicists emphasized the need for a proof of the physically plausible fact that the asymptotic distribution of the eigenvalues of a membrane of given area, or of an elastic body or of radiation in a Hohlraum [blackbody] of given volume, is independent of the shape of the area or volume; a mathematically satisfactory foundation for statistical thermodynamics seemed to depend on this theorem.[2] Characterizing the successive eigenvalues in a non-recursive manner by a minimax principle, I succeeded in proving precisely that theorem. I still remember vividly the night in my Göttingen Studentenbude when this idea came to me, seemingly without effort, and how greatly surprised I was that it actually

worked. In the meantime the kerosene lamp on my desk had started to smoke and the soot came down in flakes from the ceiling.

[6]   A number of years later, in Zurich, a pupil of mine, F. Peter, and I applied integral equations to the construction of a complete set of inequivalent irreducible representations of a compact Lie group.[3] From this work I should like to draw another lesson. The task was to carry known results over from finite to continuous groups. Molien's old paper on algebras, published in 1893, served us as a springboard; his treatment is ingenious, but clumsy according to modern standards. Had I known more about the algebraically more polished investigations of Frobenius, I. Schur, Wedderburn, etc., the going might have been harder, not easier. An awkward method sometimes contains the seeds for important generalizations, wanting in the smoother varieties. Our work had just been completed when Harald Bohr visited Zurich and gave an inspiring talk on almost periodic functions. It did not require much imagination to see the connection between his basic completeness relation for almost periodic functions and ours for the representations of groups.

[7]   More recently, integral equations have furnished the tool for proving the main proposition in Hodge's beautiful theory of harmonic integrals.[4]

[8]   The theory of integral equations has modified the face of analysis to an appreciable degree, and in view of this fact and of the impulse it gave to my own research I count the emergence of this theory with its far-flung applications as one of the outstanding, if not as *the* outstanding, mathematical event of my lifetime.

[9]   There have been others. Between 1907 and 1910 Koebe was a dominant figure in Göttingen. The *uniformization theorem* in all its various forms, which he and Poincaré first proved at that time, occupies a central position in the theory of analytic functions. Koebe never tired of composing new variations on this theme. In a limited field he had great powers of intuition; he was a constructive geometer of the first water. Klein and Poincaré, more than twenty years before, had tried to prove the theorem for the special case of algebraic functions by the so-called continuity method. Later the Riemann problem had been attacked by the same procedure, with equally unsatisfactory results. Abandoning this unwieldy instrument, Koebe and Poincaré now reached their goal by combining H. A. Schwarz's idea of the universal covering surface with simple estimates of the Harnack type for harmonic functions.

[10]   In my early Göttingen years I met Brouwer, I mean Egbertus Brouwer, the topologist and intuitionist. It was at this time (1911) that he published the series of his fundamental papers in *Mathematische Annalen*, vols. 70 and 71, proving, by means of his method of simplicial approximation, the invariance of dimension, the basic theorems about fixed points and about the degree of a topological mapping. I consider this series the second great impulse for the development of present-day *topology* after the first that originated from Poincaré's famous six memoirs in 1895–1904. Of all mathematicians I have met, Brouwer more than anybody else with the exception of Hilbert, impressed me as a man of genius. The imprint of his topological ideas is clearly visible in my book on Riemann surfaces (1913). Indeed, he and Koebe were its godfathers,—a strange couple when I now think of them, Koebe the rustic, and Brouwer the mystic. Koebe at that time used to define the notion of Riemann surface by a peculiar gesture of his hands; when I lectured on the subject I felt the need for a more dignified definition. I used the idea of cohomology for establishing the invariance of genus. Topology was in an innocent stage, then. Symptomatic for this early stage is also the fact that when Veblen and I, independently of each other, skinned Poincaré's Analysis Situs to a purely combinatorial skeleton, we did not dare to publish our investigations for several years. Topology, then a little mountain stream, has now widened into a broad rolling river. Many tributaries have flowed into it. Listing them here and now would mean to carry owls to Athens. The river has flowed far beyond my ken. But if everything is told, I still consider Brouwer's brilliant start in 1911 as the outstanding topological event of my life.

[11]   The next exciting event that comes to my mind is the solution of *Waring's problem*, presenting all integers as sums of a limited number of kth powers of integers, k being a given exponent.[5] I was present when Hilbert in the first session of his mathematical seminar after Minkowski's death outlined his proof. More intimate is my connection with the great work of Hardy and Littlewood in this field, and that of Partitio Numerorum in general. I got involved in it because my investigations on equidistribution modulo 1 provided a link the lack of which had held the work up for a considerable time.[6] The simplest case of equidistribution modulo 1, that of the multiples of an irrational number, had come my way in 1910 when I studied a very special question in heat conduction. But only when three years later Felix Bernstein told me about the application Bohl had

made of it to the astronomical problem of mean motion did I start to look into the matter in earnest. When I reported on my investigation in the Göttingen Mathematical Club, Runge ridiculed my little drawings that illustrated equidistribution of points over a square by a mere dozen of points, while Harald Bohr overwhelmed me with his knowledge of literature from Kronecker on, of which I had been completely unaware. I had laid plans to proceed from linear functions to polynomials of higher degree, but shelved them, discouraged by these criticisms. There were distractions of another kind. After my marriage and transfer to Zurich, a big paper by Hardy and Littlewood on this subject appeared in the *Acta Mathematica*. I skipped through it in the reading room of the Technische Hochschule, walked out to the terrace in front of the building where one has a marvelous view over town and lake, sat down on a bench, and in a few minutes carried out the plan conceived many months before in Göttingen, finding to my satisfaction that it clicked. I could thus prove some of Hardy and Littlewood's conjectures. Of the bearing of my results on Waring's problem I had no idea, neither then nor when I published the result in more detail after I had returned from the war in 1916. Great progress in this line of research was later made by Vinogradoff, who by more refined methods obtained estimates of unexpected accuracy.

[12]   I shall now mention by name only some major achievements of our science that in their time deeply impressed me. Personal relationships were often a contributing cause for my attention. I am a passive nature and have always been happier to learn than to think for myself. Thus passed before my eyes: Harald Bohr's theory of almost periodic functions (which was connected for me with integral equations and mean motion). The Bieberbach-Frobenius theory of the crystallographic groups in n dimensions. Hecke's analytic continuation and functional equation of the zeta function of an arbitrary algebraic field. The development of class field theory by Furtwängler, Takagi, Hasse, Artin, Chevalley, etc. (The greatest step had been made by Hilbert before I came to Göttingen as a student.) Cartan's thesis on Lie groups had been written before the turn of the century, and the main body of Frobenius' work on representations of groups is not much later. But Cartan's infinitesimal determination of the irreducible representations of all semi-simple groups and I. Schur's integral approach to the orthogonal group belong to the period under review and have influenced me profoundly. I witnessed the rise of non-commutative

algebra and of the axiomatic viewpoint in algebra to its present position, under the aegis of Schur, Wedderburn, Emmy Noether, Artin and others. In 1930 Siegel's memoir in the *Berliner Abhandlungen* that for the first time developed a systematic method for transcendency proofs made a big splash.[7] Let this be enough, though many more titles are on my lips; even in Homer's epic the ship's catalogue makes dull reading.

[13]   Einstein's "*Grundlage der allgemeinen Relativitätstheorie*" [Foundation of General Relativity Theory], published in 1916, announced a truly epochal event, the reverberations of which extended far beyond the confines of mathematics. It also made an epoch in my own scientific life. In 1916 I had been discharged from the German army and returned to my job in Switzerland. My mathematical mind was as blank as any veteran's and I did not know what to do. I began to study algebraic surfaces; but before I had gotten far, Einstein's memoir came into my hands and set me afire. You all know the influence general relativity had on the development of infinitesimal geometry. The idea for my unified field theory of gravitation and electromagnetism based on the principle of gauge invariance arose from a conversation with Willy Scherrer, then a young student of mathematics. I had explained to him that vectors when carried around by parallel displacement may return to their starting point in changed direction. And he asked "Also with changed length?" Of course I gave him the orthodox answer at that moment, but in my bosom gnawed the doubt. Mie's field conception of matter provided the ferment.[8] Once more Pythagoras' and Kepler's music of the spheres seemed to descend from heaven. Over a paper of mine on the uniqueness of the Pythagorean metric I wrote as a motto Kepler's words: "*Credo spatioso numen in orbe.*" [I believe in the geometrical order of the cosmos.] But I was not the only one who believed himself in all earnest to be on *the* road to the universal law of nature.

[14]   I have wasted much time and effort on physical and philosophical speculations, but I do not regret it. I guess I needed them as a kind of intellectual mediation between the luminous ether of mathematics and the dark depths of human existence. While, according to Kierkegaard, religion speaks of "what concerns me unconditionally," pure mathematics may be said to speak of what is of no concern whatever to man. It is a tragic and strange fact, a superb malice of the Creator, that man's mind is so immensely better suited for handling what is irrelevant than what is relevant to him. I do not share the scorn of many creative scientists and artists

toward the reflecting philosopher. Good craftsmanship and efficiency are great virtues, but they are not everything. In all intellectual endeavors both things are essential: the deed, the actual construction, on the one side; the reflection on what it means, on the other. Creative construction is always in danger of losing its way, reflection in danger of losing its substance.

[15]   In the intervals between the brain tortures of mathematical problems we must seek somehow to regain contact with the world as a whole. The probing of the foundations of mathematics during the last decades seems to favor a realistic conception of mathematics: its ultimate justification lies in its being a part of the theoretical construction of the one real world.

[16]   As for my unified field theory, the further development of physics has shown that I was right in assuming a principle of gauge invariance standing in the same relation to the conservation of electric charge as the invariance with respect to coordinate transformations stands to the conservation of energy and momentum. I was wrong in assuming that it connects the electromagnetic potentials with the gravitational g_{ij}; it rather connects them with the Schrödinger-Dirac ψs of electronic waves. But that could come to light only after the *new quantum theory* had been born. The birth of quantum mechanics is without doubt the most consequential physical event of the twentieth century; in view of its repercussions in mathematics, e.g., in the theory of operators, it is also an outstanding mathematical event. This birth is another of the great mathematical events that touched my own life. From close quarters I watched Schrödinger, who was my neighbor in Zurich, wrestle with his conception of wave mechanics. For myself, quantum mechanics became intimately interwoven with groups.

[17]   Last but not least, I have seen our ideas about the foundations of mathematics undergo a profound change in my lifetime. Russell, Brouwer, Hilbert, Gödel. I grew up a stern Cantorian dogmatist. Of Russell I had hardly heard when I broke away from Cantor's paradise; trained in a classical *gymnasium*, I could read Greek but not English.[9] During a short vacation spent with Brouwer, I fell under the spell of his personality and ideas and became an apostle of his intuitionism. Then followed Hilbert's heroic attempt, through a consistent formalization "die Grundlagenfragen einfürallemal aus der Welt zu schaffen" [to answer the fundamental questions of the world once and for all], and then Gödel's great discoveries. Move and countermove. No final solution is in sight.

[18]   I am surprised how little this serious crisis has disturbed the mind of the average mathematician. What makes him so confident? How does he know that he builds on solid rock and does not merely pile clouds on clouds? In an example like that of solving an algebraic equation one can see that the demands of an intuitionistic or constructive approach coincide completely with the demands of the computer whose task it is actually to compute the roots—with an accuracy that can be increased indefinitely when the coefficients become known with ever greater accuracy. When I preached intuitionism Landau once said: "If Weyl really believes what he says, he ought to quit his job." I did not draw such drastic consequences, but the intuitionistic attitude influenced me to the extent that I directed my interest to fields I considered comparatively safe. Of course we cannot suspend all work and just wait, hands in lap, until the underground difficulties have been settled for good. What we can do, however, is to proceed with caution and put more emphasis on explicit construction. It is not a matter of black and white, but of grades. In his oration in honor of Dirichlet, Minkowski spoke of the true Dirichlet principle, to face problems with a minimum of blind calculation, a maximum of seeing thought. I find the present state of mathematics, that has arisen by going full steam ahead under this slogan, so alarming that I propose another principle: *Whenever you can settle a question by explicit construction, be not satisfied with purely existential arguments.* One can certainly not dispense with such general notions as, e.g., that of a Riemann surface.[10] But I prefer to look upon the general propositions about them and their alleged proofs, not as statements of facts but rather as instructions for procedure in broad outlines. I would not apply the theorems mechanically to a special case but would, following the instructions, go through all the steps of proof *in concreto*, and while checking them, make them as direct, economic, and constructive as possible. I recommend this attitude as an antidote to our present indulgence in boundless abstraction.

[19]   From recording, I have inadvertently passed to criticizing. I hope I have not spoiled thereby the impression my account was intended to convey to you: It has been good to be a mathematician, and it has been a rich harvest that our science has brought in, during the last forty-odd years. I am happy that a few of its ears grew on my acre. Today we cannot help being beset by doubts in our scale of values. Yet I confess: I believe that mathematics, together with language and music, is the chief creative

reaction of man to his universe, deeply founded in his nature. I still feel the deepest satisfaction before such marvels and sublime structures of ideas as, say, the algebraic theory of numbers. The blending of constructive and axiomatic procedures seems to me one of the most characteristic and attractive features of present-day mathematics. I would not miss the axiomatic component. And yet I wonder whether we have not overplayed the game and in the perpetual tension between the concrete and the abstract have not leaned too heavily on the latter side. The mathematician is in greater danger than the physicist of following the line of least resistance; for he is less controlled by reality. A helpful guide in judging mathematical production is the distinction that Pólya and Szegő made in the preface of their famous *Aufgaben und Lehrsätze* between cheap and valuable generalization, generalization by dilution and by condensation. Take this situation. On a certain level of generality A which I call the ground level, you have certain theorems that have been proved and certain unsolved problems P of recognized interest. Suppose you discover a generalization of one of these theorems and thereby rise to a higher level of generality A'. Write it up, but lock it away in a drawer—unless or until it serves to solve one of the problems P on the ground level. At a conference in Bern in 1931 I said: "Before one can generalize, formalize, and axiomatize, there must be a mathematical substance. I am afraid that the mathematical substance in the formalization of which we have exercised our powers in the last two decades shows signs of nearing exhaustion. Thus I foresee that the coming generation will have a hard lot in mathematics." That challenge, I am afraid, has only partially been met in the intervening fifteen years. There were plenty of encouraging signs in this conference. But the deeper one drives the spade the harder the digging gets; maybe it has become too hard for us unless we are not given some outside help, be it even by such devilish devices as high-speed computing machines. Should we thus seek a broader contact with reality and other fields of knowledge? That seems to be the trend among a considerable section of the younger set. It is to be welcomed. For even from a purely philosophical standpoint, the conception that mathematics is essentially a part of the theoretical construction of the one real world is in better accord with our probings of the foundations of mathematics than more idealistic views.

[20]   Mathematical papers are read by but a few people. How many more can enjoy a work of music! If the aesthetic side of mathematics is essential,

our endeavors look somewhat futile. I think the main purpose of the day-by-day work in mathematics is to create the atmosphere in which the work of genius can thrive. When I compare the present average standard of mathematical papers with the past, I find that it has risen. I do not think that such a flood of mediocre papers as were written in the heyday of integral equations could now pass the gates of our journals. But the mass of little results that claim attention has made it harder to accomplish the exceptional; hence the net effect is a leveling one. What has definitely deteriorated is the art of writing along with the deterioration of modern languages in general. Writers of fifty years ago, fewer in number, tried harder to make their points clear to a wider group. Basic ideas, connections with other fields, formulated in a not too technical language, received more emphasis. Perhaps our predecessors were aided by a more solemn conviction of the importance of their scientific effort. A large percentage of our papers, if they are not outright monologues, exclude by their very style whoever does not belong to the narrow circle of people working along exactly the same lines. Research conversation between members of such a group should be distinguished from the communication of significant developments to a wider public. Not the first but the second is the proper purpose of publication. Maybe the first should be carried on in another way than by printed papers, say by classified or unclassified reports, and only communications of the second type should go into print.

[21]   Ever since Newton, in his modesty, spoke of playing with pebbles on the shore of a wide ocean, the attitude that we mathematicians play a nice game that ought not to be taken too seriously has enjoyed considerable popularity. In my opinion it is fundamentally unsound. Whatever analogies there are between the mental activities of a mathematician and a chess player, the problems of the former are serious in the sense that they are bound up with truth, truth about the world that is and truth about our existence in the world. That the game of mathematics and physics is a harmless game, one of Hardy's main points in defense of mathematics, nobody can claim any longer. Hardy distinguishes between the real mathematics, that of Gauss and Riemann and Ramanujan, and the dull Hogben type of Mathematics for the Million, and thinks that only the latter is useful or harmful as the case may be.[11] I do not believe that his distinction is valid. Riemann is as respectable a mathematician in his papers on the hydrodynamics of shock waves as in his paper on the zeta

function. If the progress of science is blamed for the impasse in which the world finds itself today, then the mathematician has to assume his share of responsibility.

[22]   We may well envy the nineteenth century for the feeling of certainty and the pathos with which it praised the sacrosanctity and supreme value of science and the mind's dispassionate quest for truth and light. We are addicted to mathematical research with no less fervor. But for us, alas! its meaning and value are questioned from the theoretical side by the critique of the foundations of mathematics, and from the practical-social side by the deadly menace of its misuse.[12]

[23]   *Crescet scientia, pereat mundus!* [Let science grow, though the world perish!] The progress of science has led the world to a terrible impasse. Can we plead innocent before the tribunal of civilization, and put all the blame on the wicked who misuse our knowledge? There is indeed much to be said in our defense: Without the gift of creative thought and its free exercise, human life would not be what it is; also this, that science is neutral towards good and evil. And yet, it is made by men for men and must not be isolated from man's total existence. Is this then our sin, that we let the unity of existence go to pieces? Or do we, in following the urge of intellectual curiosity that is deeply implanted in our nature, merely fulfill our destiny, like the silkworm?:

Verbiete du dem Seidenwurm zu spinnen,
Der sich dem Tode immer näher spinnt.

[Forbid a silkworm to weave. No use.
He will weave on, though weaving his own death.][13]

[24]   Some people think we can be saved by striking a better balance between the social and the natural sciences. With all respect for the social sciences, but this advice demands too much of them! Technological knowledge is such a dangerous tool in the hands of man, because of the second law of thermodynamics; it is much easier to blow up a building than to build it. What would really be needed to offset the menace of the progress of natural science is a development, not of social *science*, but of social *behavior* and moral responsibility, of our whole attitude towards life. But alas! they change much more slowly than our knowledge. Here is the frightful dilemma that may spell our doom.

[25]   Thus the meaning and value of mathematical research is questioned, both from the theoretical side by the critique of the foundations of mathematics, and from the practical-social side by its dreadful implications. What to do is a question everyone of us must answer according to his own conscience. I can suggest no universal solution.

7 ▣

Man and the Foundations of Science

ca. 1949

One does not miss the mark badly by fixing as the birth date of our science and our philosophy that day when Democritus spoke the following words ringing down the ages: "Sweet and bitter; cold and warm, as well as the colors, all these things exist but in opinion and not in reality (*nomōi, ou physei*); what really exist are unchangeable particles, atoms, and their motions in empty space." Indeed, reflecting philosophy is impossible as long as the standpoint of naive realism stands unquestioned, constructive science is impossible as long as one accepts all phenomena at their face value. The rock on which science is built, according to Democritus, is a conglomerate of the fundamental concepts of space, time, and substance or matter.

Time as a one-dimensional manifold is a relatively trivial affair for the constructive scientist; *space* with its three dimensions is not so easily disposed of—geometry is an elaborate doctrine. By its precision and certainty it far exceeds any other knowledge of the external world, and its evidence seems not to depend on any special experiences such as teach us that fire burns and spring returns after about 360 days. "*Nihil veri habemus in nostra scientia nisi nostram mathematicam* [We have nothing true in our science except our mathematics]," says Nicolaus Cusanus [Nicholas of Cusa], and Kepler: "The science of space is unique and eternal and is reflected out of the spirit of God. The fact that man may partake of it is one of the reasons why man is called the image of God." From Galileo I quote two relevant passages: "It is true that the divine intellect cognizes the mathematical truths in infinitely greater plenitude than does our own (for he knows them all), but of the few that the human intellect may

175

grasp, I believe that their cognition equals that of the divine intellect as regards objective certainty, since man attains the insight into their necessity, beyond which there can be no higher degree of certainty." And the other famous passage where he first attacks the speculative metaphysicists who treat philosophy like a product of pure imagination, such as the *Iliad* or *Orlando Furioso*, in which it is of little importance whether what is said is true, and then continues: "But the true philosophy is written in the great book of nature which is continually open before our eyes, but which no one can read unless he has mastered the code in which it is composed, that is, the mathematical figures and the necessary relations between them."

In his grandiose abstraction from sensual semblance Democritus reduces all variety of the world to the absolute contrast of the void and the full (*mē on* and *pamplēres on*) which has to be accepted as an irreducible principle. His *pamplēres on* is incapable of any qualitative differences. Now the concept of motion depends on the possibility to identify the same substantial particle at different times and ascertain its positions in space. For a homogeneous substance this is impossible unless the substance consists of isolated particles. They move, but a motion within an individual particle is indiscernible, and hence they are atoms of unalterable shape. Specific gravity of a macroscopic body results from the proportions of volumes in which atoms and empty interstices are mixed. Gassendi renews the atomic theory at the beginning of the seventeenth century. Here is Galileo's echo: "The variety exhibited by a body in its appearances is based on dislocation of its parts without any gains or losses. Matter is unchangeable and always the same because it represents an eternal and necessary form of being." An atom is indivisible and rigid; i.e., the space covered by it remains at all times congruent to itself. The parts of space occupied by various atoms have no points in common, or the atoms are impenetrable. *Solidity* is the word used by Gassendi to indicate both rigidity and impenetrability. It is to be conceived as an abstract geometric property; it neither depends on the experience of the subjective quality of hardness, nor should it be interpreted dynamically as the result of forces holding the atom together. Huyghens, who thinks in terms of spatial mechanical principles, makes this very clear in a controversy with the metaphysical dynamist Leibniz.

Is it necessary to hear more witnesses about the subjectivity of sensual qualities and the objectivity of geometric ideas? This is Locke's distinction of secondary and primary qualities. Says he: "The idea of primary qualities of bodies are resemblances of them, and their patterns do really exist in the bodies themselves; but the ideas produced in us by the secondary qualities have no resemblance of them at all. . . . What is sweet, blue or warm in idea, is but the certain bulk, figure and motion of the insensible parts in the bodies themselves, which we call so." Or even more forcibly: "A piece of manna of a sensible bulk, is able to produce in us the idea of a round or square figure; and by being removed from one place to another, the idea of motion. This idea of motion represents it, as it really is in the manna moving; a circle or square are the same, whether in idea or existence; in the mind or in the manna. And this, both motion and figure are really in the manna, whether we take notice of them or not. This everybody is ready to agree to. Besides, manna by the bulk, figure, texture, and motion of its parts has a power to produce the sensation of sickness, or sometimes of acute pains or gripings in us. That these ideas of sickness and pain are not in the manna, but effects of its operations on us, and are nowhere when we feel them not: this also everyone readily agrees to. And yet men are hardly to be brought to think, that sweetness and whiteness are not really in manna; which are but the effects of the operations of manna, by the motion, size and figure of its particles on the eyes and palate; as the pain and sickness caused by manna are confessedly nothing but the effects of its operations on the stomach and guts, by the size, motion and figure of its insensible parts."

This epistemological doctrine is, in the seventeenth and eighteenth centuries just as at Democritus's time, closely related to the mechanistic conception of the universe. Listen to Newton, the incorrigible theist, in the concluding remarks of his *Opticks*: "All these things being consider'd, it seems probable to me, that God in the Beginning form'd Matter in solid, massy, hard, impenetrable, movable Particles, of such Sizes and Figures, and with such other Properties, and in such Proportion to Space, as most conduced to the End for which he form'd them; and that these primitive Particles are Solids, and incomparably harder than any porous Bodies compounded of them; even so very hard, as never to wear or break in pieces; no ordinary Power being able to divide what God himself made one in the first Creation. . . . And therefore, that Nature may be lasting, the changes of corporeal things are to be placed only in the various Separations and Motions of these permanent Particles." Besides the theistic there is in this description a slight dynamical undertone, and indeed a few lines later, Newton continues: "It seems to me farther, that these particles have not only a *Vis inertiae*, accompanied with such passive Laws of motion as naturally result from that Force, but also that they are moved by certain active

Principles, such as is that of Gravity, arid that which causes Fermentation, and the Cohesion of Bodies." This goes far beyond Democritus's original simple conception. Therefore I repeat its essentials in a less baroque language which stresses the role of space in this world picture: Space is an infinite manifold of points, and yet homogeneous. Hence only by severing the full and the void, the parts of space covered and uncovered by matter, can this diversity of points result in concrete patterns describable in geometric terms. This pattern of atoms can, and will in general, change in time. It is thus the latent diversity of points in space, made patent by the contrast of void and full which lies at the bottom of the ever-changing aspect of the world around us.

By this picture the study of nature is reduced to two questions: first, of what shape are the atoms which build up matter? Second, what are the laws of motion? The simplest hypothesis about the first point ascribes spherical shape to the atoms; for then the motion of an atom is reducible to the motion of a single point, its center. Under these circumstances various atoms can differ by their radii only. (More fanciful writers depict atoms with hooks which are hooked together when they form a solid chunk of matter. How one who breaks the chunk unhooks all these little (unbreakable!) hooks is beyond my imagination.) Concerning motion, Democritus himself naturally assumes that the atoms as long as they do not collide fall freely in space from top to bottom. Since Galileo, this unperturbed motion is, of course, the uniform motion in a straight line with constant velocity. When two atoms collide they change their motion abruptly according to laws which were correctly formulated by Huyghens; they are what we now call the laws of conservation of energy and momentum, supplemented by the assertion that no [net] exchange of momentum occurs in the directions which lie in the common tangent plane of the colliding bodies at their point of touch.[1] True, there enters into these laws a certain coefficient associated with each particle, called its mass. But for the consistent mechanist, mass is the quantity of matter as measured by its volume.

In Galileo's, Newton's, and Huyghens's physics, all occurrences are constructed as motions in a space which is at the same time intuitive and objective. Huyghens, who, as you all know, developed the undulatory theory of light, can say with the best conscience that colored light beams are in reality oscillations, waves, of an etherial fluid which probably consists of especially tiny bits of particles. Let us not underrate the simplicity, precision and persuasive power of this world picture! Yet how had one arrived at it? Out of the complete reality, dressed as it were in the glittering adornment of qualities, a few features, the geometric kinematical ones, have been picked out. For some reasons which today we find hard to realize, they were deemed to be particularly trustworthy, and with these elements one built up a true objective world in the face of which the world of our daily experience fades to a mere semblance. From this standpoint a civilized man should almost feel ashamed to stoop to this confused world of shadows, as he still does in the conduct of his daily life. One who is reluctant to sacrifice to the natural sciences all other human endeavors moving in the medium of words, has no other way out than to resort to the dubious doctrine of the splitting of truth into several views of the world, each closed in itself. It is hard to believe how stubbornly philosophers and physicists alike stuck to this interpretation of physics, even after physics itself in its factual structure had long outgrown it. Their excuse is that of an ocean traveler who distrusts the bottomless sea and therefore clings to the view of the disappearing coast as long as there is in sight no other coast toward which he moves. I shall now try to describe the journey on which the old coast has long since vanished below the horizon. There is no use in staring in that direction any longer. I have the impression that recent developments in mathematics, physics, and philosophy, in spite of their almost complete independence, begin to converge toward a new viewpoint. A new coast seems dimly discernible, to which I can point by dim words only, and maybe it is merely a bank of fog that deceives me.

Democritus's old scheme was punctured in so many ways that I am at a loss where to begin. First let us look at epistemology. Objectivity of space and time became as suspect as that of the sensual qualities. Already Descartes had felt the need for justifying the position. He is as strong as anybody else in rubbing it in that there is no more resemblance between an occurrence and its perception (sound-wave and tone, for instance) than between a thing and its name. It is different with the ideas concerning space, because in contrast to the sensual qualities we recognize them clearly and distinctly, and a fundamental principle of his epistemology claims truth for whatever one knows in such a way. In order to support that principle he appeals to the veracity of God, who is not bent on deceiving us. Descartes had grasped the fundamental tenet of idealism that nothing else is given than the immediate data of my consciousness. The problem is how to reach out from there toward a transcendental real world common to all men, in which my consciousness assumes the role of a real individual, beside innumerable others who claim equal rights. Obviously one cannot do without a veracious God guaranteeing truth, if in spite of one's idealistic conviction one builds up the real world out of certain elements of consciousness, for instance spatial intuition, that for some reason or other seem particularly trustworthy. Two hundred years later Georg Büchner, the German revolutionary and playwright who wrote *Danton's Death* and *Wozzeck*, mocks at Descartes: "How keenly he measured out the grave of philosophy! His use of the dear God as the ladder to climb out of it is, to be sure, strange. The attempt turned out somewhat naively, and even his contemporaries did not let him get over the edge." The encyclopedist d'Alembert no longer justifies the use of spatial-temporal notions as constructive material by their clarity and distinctness as Descartes did, but simply by the practical success of the method.

Leibniz was the first to declare quite explicitly: "Concerning the bodies, I am able to prove that not only light, color, heat and the like, but motion, shape, and extension also are merely apparent qualities." The doctrine of the ideality of space and time is the cornerstone of Kant's transcendental idealism. He describes space and time as forms of our intuition: "That in which sensations are merely arranged, and by which they are susceptible of assuming a certain form, cannot be itself sensation; hence, indeed, the matter of all phenomena is given to us only a posteriori (namely by means of sensations), but the form of them must lie ready a priori within our mind and therefore must be capable of being considered independently of all sensations." And this explains for him the a priori certainty of the science of space, Euclidean geometry. Space as form of my intuition can scarcely he described more suggestively than by these words of Fichte: "Translucent penetrable space, pervious to sight and thrust, the purest image of my awareness, is not seen but intuited and in it my seeing is intuited. The light is not without but within me, and I myself am the light."[2]

For the same reason as sense qualities were expunged from our objective world picture, we are now forced to eliminate even space and time. The device by which to accomplish this had been invented by Descartes; it is analytic geometry. After choosing a definite system of coordinates in the plane, any point can be determined and represented by its two coordinates x, y, i.e., by a pure numerical symbol (x, y). The circumstance that several points (x, y) lie on a straight line is translated into the assertion that their number symbols (x, y) all satisfy a certain linear equation. The equality of two distances $AB, A'B'$ is expressed as a simple arithmetical relation between the coordinates of the points A, B, A', B'; and so on. We thus construct what one may call an arithmetic model of geometry. The whole of geometry has been so thoroughly arithmetized that in our day the mathematician always proves his geometric theorems for n-dimensional space, without taking any particular interest in the case $n = 3$ which our actual space realizes, and nobody would be able to draw a clear line between geometry on the one hand, algebra and analysis on the other. The same in physics. Instead of maintaining that the objective world itself consists of solid particles moving in space, it would have been more modest to claim that such a construction can be used as a *model* for the real world, from which one can read what happens in reality by carrying out the corresponding constructions in the model. This is how the designing engineer or architect proceeds, who carries out his geometric constructions on paper. But in theoretical physics this procedure has long since and almost completely given way to analytic calculations in which the points in space and moments in time appear in the form of their coordinates. A quantity like the temperature of a body, or the electric field strength in an electric field, which has a definite value at each space-time point, appears as a function of four variables, the space-time coordinates.

While the bottom of space and time thus dropped out of the barrel, all sorts of other notions crept in. The trouble started right at the beginning with Galileo's introduction of inertial mass. When one takes down the scaffolding that helped him to climb up step by step to this idea, one is left with the following explanation. As long as a body is not influenced from outside, and hence moves in a straight line with uniform velocity v, it possesses a certain impetus or momentum J which is a vector parallel with the velocity v. The factor by which one has to multiply v in order to get the momentum J is the mass. Hence we should know what mass is as soon as we are told what momentum is. To this question Galileo answers by a law of nature rather than by an explicit definition, to wit, the law of momentum: If several bodies enter into a reaction, the sum of their momenta after the reaction, when they have ceased again to influence each other, is the same as before. When one subjects the observation of reacting bodies to this law, introducing if necessary auxiliary bodies, one is able to determine the relative values of their masses. In order to measure the mass of a body one must therefore have it react with other bodies. Inertial mass is a *concealed* character; we cannot perceive it directly as we can its color. Moreover, the determination of mass is possible only on the basis of a law of nature to which that notion is bound. The laws of nature appear half as expressions of fact, half as postulates. Finally, we attribute a mass to each body whether or not we actually perform the reactions necessary for its measurement; we rely upon the mere *possibility* of carrying them out.

With this introduction of mass, a step of immense consequence had been taken. After matter had been stripped of all sensual qualities, science at first seemed restricted to the use of spatial attributes only. But now we discover that by letting a body react with others we may ascertain numerical characteristics of a different type, like mass. Besides carrying out the reactions one must submit them to a definite theory (here consisting of the law of momentum). The implicit definition and the experimental measurement of such a characteristic are based on theory. It is this procedure which opens the sphere of mechanical and physical concepts proper beyond geometry and kinematics.

It could be proved that a swarm of Huyghens's spherical atoms obeying his laws of collision behaves like a *gas*. His theory therefore could neither explain the fluid nor the solid state of matter. But even for gases it was unsuccessful at a crucial point. In comparing the theory with observations, one obtained fairly reliable values for radius and mass of the atoms. But it proved untrue that for the several chemical elements the atomic masses are proportional to their volumes. This result demolishes the conception of a uniform dough of substance from which the Creator at the beginning of all time cut out the little cakes of atoms, baked them to absolute rigidity, and let them loose with various inertial velocities. However, the masses derived from the kinetic theory of gasses are in agreement with the relative atomic weights at which chemistry arrived by abstracting its law of multiple proportions from a vast material. For a long time chemistry alone could give substantial empirical support to atomism.

While a body K is under the influence of other bodies, say k_1, k_2, k_3, its momentum in general will change. The rate of change per unit of time measures the influence or the *force* exerted on K by k_1, k_2, k_3. Indeed, in studying the motion of the planets, Newton finds that this force is the sum of three individual forces ascribable to k_1, k_2, k_3, individually, so that, for instance, the force of k_1 upon K depends only on the state of the bodies K and k_1, and not on k_2 and k_3. More explicitly, the force acting between two bodies will depend on their positions, velocities, and also on their inner states. The latter will enter into the law of the force by certain characteristics of that inner state, as for instance the electric charge into Coulomb's law of electrostatic attraction and repulsion. Thus the notion of force becomes a source of new measurable characteristics of matter.

Hand in hand with these developments the idea of matter becomes more dynamic than substantial in nature. Hardness and impenetrability give way to repulsive forces, cohesion is effected not by hooks, but by attractive forces. The atoms become "centers of force." Boscovich, Cauchy, and Ampère assume that the centers are points without extension. Kant constructs matter by an equilibrium of attractive and repulsive forces. The old geometric-mechanistic picture gradually gives way to the physics of central forces.

The next decisive step is the replacement of electrostatic force acting into distance by continuous propagation of an electromagnetic field. Maxwell made it almost certain that light consists of high-frequency oscillations of this field. The question what the electromagnetic field is can be answered as little as the question what mass is. But here are the facts. In the space between charged conductors, a small charged "test body" at a point P experiences a certain force H; the same force again and again whenever I bring the test body into the same position P. In varying the test body one realizes that the force H depends on it, but in such a way that one can split H into two factors $e \cdot F$, the first [e] being a scalar depending only on the state of the test body, and neither on the point P nor on the state and position of the conductors, while the other factor F, a vector function of the point P, is in its turn independent of the test body. e is the charge of the test body; the vector F is the field strength at P. The test body is here used to measure the electric field.

But it belongs to nature no less than the conductors generating the field. Hence in the course of our theory a law must be established giving the pondero-motoric reaction of the field upon the charged bodies by which it is generated. This law may or may not agree with the definition from which we started. Thus we had better not commit ourselves to any definition and rather develop the theory as a symbolic construction with unexplained symbols and only at the end indicate in which way certain derived quantities may be checked by observation. The theory then becomes a connected system that only as a whole may be confronted with experience. I take the theory of electromagnetic phenomena as an example, but make allowance for two simplifications: I assume instantaneous propagation rather than propagation with finite velocity (although the finite velocity is the most distinctive feature of Maxwell's theory, which makes it as superior to the physics of central forces as Newton's dynamical theory of gravitation was superior to Kepler's geometrical description of the planetary orbits). Secondly I assume that we deal with one sort of particles, "electrons," all endowed with the same mass and charge, and that it is their motion which is directly observable (an almost unpardonable idealization!). Knowing the positions and the velocities of the electrons at a certain moment t, we want to ascertain them at an immediately following moment $t + dt$. If this is possible, we can derive the whole motion of the particles from moment to moment, and compare the theoretical result with our observations. According to certain laws, the given positions and velocities at t uniquely determine the electromagnetic field at the same moment. By further laws the field determines the distribution of energy and momentum in space; and by means of the flux of momentum the field exercises certain ponderomotoric forces upon the generating particles. Finally, force imparts a certain acceleration according to Newton's fundamental law of motion. Velocity and acceleration are the rates of change of position and velocity respectively, hence they allow us to compute the change of position and velocity taking place during the following little time interval dt. Only this entire connected theory, into the texture of which geometry also is interwoven, is capable of being checked by observation. The situation becomes considerably more complex if we abandon our unpardonable idealizations. An individual law isolated from this theoretical structure simply hangs in the air. Ultimately all parts of physics including geometry coalesce into an indissoluble unit.

At each step of this construction, penetrating into ever deeper layers, the physicist tries to invent an intuitive language which deals with the new entities as if they were such familiar objects as table and bed. But in the systematic theory one should skip all intermediate stops, put down the symbolic construction without explaining what mass, charge, field strength, etc., mean, and then try to describe how this structure ties up with our immediate experiences. It is sure that on the symbolic side not space and time, but four independent variables will occur; it is on the side of consciousness that we shall have to talk about intuitive space perceptions, for instance coincidences in space, just as about sounds and colors. Of a monochromatic beam of light, which with Huyghens was in reality an oscillation of the ether, we now should have to say that it is a mathematical formula expressing a certain symbol F called electromagnetic field strength in terms of other symbols t, x, y, z, called time-space coordinates. It becomes evident that now the words "in reality" must be put between quotation marks; we have a symbolic construction, but nothing which we could seriously pretend to be the true real world.

First Maxwell himself tried to design mechanical models which would explain his laws of the electromagnetic field. Such attempts continued, but were gradually abandoned during the second half of the nineteenth century. If one defines precisely enough what a mechanical model is, one can even prove the impossibility of the task. But such a showdown is hardly needed. The decisions of history in such cases are infallible and irrevocable. The mechanistic world picture is dead, beyond restoration. For a while one was inclined to go to the other extreme and to assume the field to be the only physical entity, conceiving the electrons and other particles as small areas in which the field strength rises to enormously high values. They would move about in the field almost like waves over the surface of the sea. Meanwhile, refined experimental technique had ever more clearly exhibited the atomic features of matter. But then quantum physics came along and revealed field or wave, and particles, as two incomplete aspects of the same situation.

In sketching the further development I shall separate mathematics and physics. In mathematics, as we have seen, space and geometry have been arithmetized. In physics the necessity of this step became most conspicuously evident by the general theory of relativity, which broke with Euclidean geometry and reduced the geometric properties of things to effects of a physical field called the gravitational field because it is also responsible for the phenomena of gravitation. Mathematics, however, did not stop at arithmetic. It endeavored to reduce arithmetic to pure logic. The primitive logical concepts "and," "or," "not," "if, then," together with the general notions of a *set* of objects and of *mapping* such sets upon each other, were to serve as the only foundations, taking over the role which space, time and matter had served in Democritus's scheme. The attempt, initiated by Dedekind and Cantor, and most consistently pursued by Bertrand Russell, was successful to a certain extent, and today the set-theoretic language permeates mathematics throughout. However, one had to pay a high price, the price of running counter to the plainest evidence concerning the infinite, and one was punished for it by actual contradictions which showed up menacingly, though only in the farthest boundaries of mathematics. For physics the most important form of the infinite is the one-dimensional range of real numbers. Somewhat simpler is the infinite sequence of integers $1, 2, 3, \ldots$, which we use for counting, and I shall point out briefly the difficulty for this case. We are not content with picking up numbers as they actually occur in our life, when we count oranges or books or dollar bills, but we create a priori the sequence of all possible integers by starting with 1 and proceeding from one number to the next. We have the feeling that we can do this again and again, beyond any point already reached. The sequence, naturally, is never completed; rather it is a field of possibilities open into infinity. Thus Being is projected onto the background of the Possible, or, more precisely, into an ordered manifold of possibilities producible according to a fixed procedure and open towards infinity. Essentially the same is done when we try to fix a point in a continuum, more and more precisely, by casting a net of partition over it, describable in purely combinatorial terms, which is refined step by step *ad indefinitum* by iteration of a universal process of subdivision. Now it is clearer than ever that space-time as the open field of possible coincidences cannot be spared the combinatorial symbolic treatment to which mathematics subjects whatever comes under its hand. By keeping open the possibility of going beyond any step reached, the mathematician is prepared to meet any refinement of the physicist's measurements. The sin committed by the set-theoretic mathematician is his treatment of the field of possibilities open into infinity as if it were a completed whole all members of which are present and can be overlooked with one glance. For those whose eyes have been opened to the problem of infinity, the majority of his statements carry no meaning. If the true aim of the mathematician is to master the infinite by finite means, he has attained it by fraud only—a gigantic fraud which, one must admit, works as beautifully as paper money.

And we are so willingly deceived by it because without it mathematics would lose 99 per cent of its power. The way out which Hilbert devised in our day is, to be frank about it: to change the meaningful statements into meaningless formulas composed of symbols. One plays with these symbols and formulas according to certain practical rules, thus deducing one formula from others, not unlike the chess player who proceeds from one position of his men to another according to the rules of his game. The process of formalization extends to both logic and mathematics. Meaning and understanding are not entirely eliminated, though. While the mathematical formulas are deprived of it, one must understand how to handle the symbols and how to play the game. The symbols like 1, +, etc., are concrete tokens drawn with chalk on blackboard or with pencil on paper. One must understand what it means to place one after the other, one must be able to recognize them in their recurrences irrespective of irrelevant details of their execution. And then one must understand how to apply the rules correctly. But in the system itself, which had been intended to represent the true objective world, even the last remnant of meaning, the last straw to which the logicists, Bertrand Russell, and the Viennese school clung for a little while, is swept away. We are left with our symbols, tokens drawn with chalk on a blackboard. With them we deal on the same footing as with other utensils in our daily life, as we open a door to enter a room, sit down in a chair, travel to a meeting, or call on a friend, and we rely on the same kind of understanding. We move in the world of our seeing, acting, caring, natural life, and by no means in the realm of the immediate sensual data of consciousness, about which the positivists used to talk so much; in a world so infinitely more obvious and familiar to every one of us, although the suspicious analyzing intellect finds it bewilderingly complex and muddy. As scientists we may be tempted to argue: As we know, the chalk on the blackboard consists of molecules, and these are swarms of charged and uncharged electrons, neutrons, etc., which ultimately dissolve again into mere symbols and formulas, which in their turn are written with chalk on a blackboard . . . You see the ridiculous circle.

One more remark illustrating the situation by a typical example before I go on to quantum physics: In formalized mathematics a symbol $\rightarrow$ occurs between formulas; e.g. $a \rightarrow b$. This formula $a \rightarrow b$ has replaced the statement that the proposition a implies b or that b *follows* from a. It recalls and points to this logical concept of implication, although now it has become a meaningless symbol. On the other hand, when we play the game of deduction and start with a certain formula a we may arrive at b we could then say: As our deduction shows, b follows from a. This "follows" has a meaning, and looking down, as it were, with our mind's eye on the game we convey by our words a meaningful statement about what has happened in the game. In practice we are careless enough to enunciate our mathematical theorems in words rather than in formulas, as if they still kept their old meaning. But in principle we must sharply distinguish between the symbol $\rightarrow$ occurring within the system, and such words as "follows" which we use to make meaningful communications about the game.

I seem to see that a split of essentially the same nature has been brought about by quantum physics: namely the split between the physical phenomenon under observation on the one hand and the measurement on the other hand. The first can be adequately described only by the quantum-mechanical symbolism; about the latter we can and must talk in the intuitive terms of classical physics. The new situation which has arisen through quantum theory has been described in various ways. We saw that the ideal characteristics which physics ascribes to bodies are based on reactions with other bodies. For instance, we measure temperature by bringing the body into contact with a thermometer. In quantum theory we learn that measuring one quantity sometimes utterly destroys the possibility of measuring another quantity. This is a matter of principle and not of human deficiency. For instance, the position and momentum of an electron are complementary quantities in this sense, and the prevailing indeterminacy is always indicated by Planck's famous quantum constant h. One speaks of an uncontrollable encroachment of measurement on the phenomenon to be measured. Another feature of quantum physics is the double aspect every phenomenon assumes in the wave and the particle picture. Moreover [the] connection between the phenomenon described in symbols and the observation of individual particles is of a statistical nature. But probably the most essential point is the one I mentioned first: the distinction between phenomenon and measurement. The former is woven of an airy stuff which we can hardly approach by means of our ordinary language; only the symbolism is its adequate representation. But the results of measurements belong to the disclosed world of our daily experience; they are describable in terms of classical physics and we can understand each other, communicate, and compare notes about them. (For every question referring to such measurements, quantum theory provides in principle an unambiguous answer.) I remember when Niels Bohr talked in Princeton about these things, he used to picture the phenomenon by very delicate and vague red lines; while the measuring apparatus, for instance a frame in which one measures a position, was drawn with thick yellow lines, indicating, as in a workshop drawing, the thick boards by shading, and not omitting the screws. This is to the point: In the same naive way as the mathematician handles his symbols (without caring what physics pretends his ink or chalk to be "in reality"), so the experimental physicist must handle his boards, screws, pipes, wires, and understand how to read a pointer, apply a yardstick, etc. Here he is on the same level of existence as the carpenter or the mechanic in his shop, on the same level of understanding on which our rough and tumble daily life moves.

What is the conclusion to which we come? Modern mathematics and physics may seem to move in thin air. But they rest on a quite manifest and familiar foundation, namely the concrete existence of man in his world. Science is not engaged in erecting a sublime, truly objective world in such pure material, as Democritus employed, above the Slough of Despond in which our daily life takes place. It simply endeavors to prolong a certain important line already laid out in the structure of our practical world. By no means does it pretend to exhaust concrete existence. For instance, the reflections with which we are occupied just now are of a different kind. Terms like "values," "social acts," "history," point in still other directions, and these lines may also admit a similar prolongation. There need not result any serious objective conflict as long as none of these directions claims to erect beyond this vulgar world of our common understanding, a higher, purified, the only true one, in terms of which all things in heaven and on earth have to be interpreted.

Recent developments in philosophy seem to me to converge toward a similar attitude. I mean the gradual transition from idealism to a kind of philosophical anthropology. The idealists, Descartes, Berkeley, Kant, Fichte, all start from a pure ego, a pure consciousness. Hegel was the first to make man in his concrete historical existence the center of his philosophy. Kierkegaard, Nietzsche, Dilthey, Scheler, follow in the same direction, and Heidegger's and Jaspers' existential philosophies for the present top the movement.

I find it extremely hard to translate the essential tenets of existential philosophy into plain English, and I shall deliberately flatten Heidegger's terribly involved phrases. As far as these tenets concern our problem they are perhaps explained best by confronting them with the idealistic doctrine.

A preliminary remark: immanent consciousness, being for oneself, mind, the realm of images, *res cogitans*, however you may call it, in contrast to *esse*, mere being, is always *my* consciousness.[3] Somebody else's consciousness *qua* consciousness is closed to me. An apparition can only be an apparition for me. Of course, I shall not prevent you from asserting the same fact, each for yourself.

Immanent consciousness is the starting point of the idealist. As the primarily given, he finds the images of my consciousness—just as I have them. If he happens to be an extremist, he will admit nothing but sensations. Suppose I perceive a green hat. Would not the unsophisticated describe the situation somewhat as follows? In this act of perception I sight green, quite apart from the question whether there is in reality a hat and whether it is really green. By reflection I may become aware of my own perception, which then is turned into the object of a secondary act of inner awareness, and only thus can the perception be analyzed, with such results as these, that it radiates from an ego and points toward an intentional object "green" or "green hat." But even without the question of reality being raised, neither for the object nor the subject, the radical idealist will doubt very much whether this ego and the intentional object are inherent parts of the given perception, just as much as he doubts that in the perception of a body the back side is given. He takes pains to describe how by a sort of productive synthesis or "psychic chemistry" the intuition of space constitutes itself in a number of steps, beginning with the two-dimensional extension of the visual field. Even this he wishes to reduce

189

to "local signs" (Lotze), sensations whose qualitative gradations correspond to the various positions in the visual field (of course he hunts in vain).[4] A serious problem is seen in the splitting of a presumed unitary sensation into color and extension; one asks in how far two sensations of the same red, localized at two different spots P, Q of the visual field, by being both red, have a closer kinship than a sensation of red at P and of green at Q. The perspective views of an object which an observer gathers in changing his standpoint constitute the three-dimensional body itself. But does not each such step of synthesizing the immediately given to form something of a higher order, violate the idealist's own creed? Does it not transcend that which is given? and what is the synthetic principle in each case? The most difficult crux is the thesis of *reality* which undoubtedly is involved in all our acts of perceiving, remembering, etc., even if only in the sense that reality is pretended by them. Indeed no single observation gives an absolute right to ascribe existence and the perceived qualities to the perceived object; it may be outweighed by other perceptions. Approach to reality is an infinite process of ever new, partly contradictory perceptions and their mutual adjustments. But the idealist is charged to justify by the immediate data of consciousness these adjustments and with them the general thesis of a real external world. I believe that every attempt to vindicate on such a basis the belief in a real outer world and in the reality expressed by the word "you" has been a failure and must be a failure. And the mode in which things are understood as implements in daily life, how we use them, our acting upon the world, simple social acts like requesting or promising, the mode in which we understand (and misunderstand) each other from man to man in our intercourse, the mode in which spoken words are carriers of meaning, the realms of art and of ethics, these and many other things are not yet even touched!

Existential philosophy turns this whole construction upside down, without relapsing into a crude realism, and thus overcomes the old contrast of realism and idealism, which had grown stale in the course of centuries. It embraces what is most valuable in pragmatism. Previously Gestalt psychology had already done away with the mosaic of sensations and shown in detail how the mind grasps a whole which transcends the sum of its parts, becoming aware of the parts only through the whole. Phenomenology had lunged out in all directions beyond the narrow boundaries of sensual data; but still its pure ego was like a spirit soaring above the waters, untainted by worldliness, and it was hard to conceive how it could ever surrender its immanence and become flesh, man among men.

The form of being which is revealed in us to ourselves is called *Dasein* by Heidegger.[5] He stresses its existence, although it is for him as for the idealist that form of being whose understanding must precede the understanding of all other forms of being, because it understands itself. He uses the term "understanding" for the *cogitare*, but it includes from the beginning the disclosedness in which I find myself in the world and meet my follow men. Therefore he says: Dasein is being whose very being implies understanding of this being. It is always myself. It understands itself as being-in-the-world and being-with. In this disclosedness it understandingly manages things (*prāgmata*) and communicates with other fellow-beings. For one who denies this manifestness every access is blocked. It is no objection that, as a matter of fact, this understanding awakens from, and sinks back into, Nirvana. Under the title of *time* it will have to account for this its feature. Only out of the fullness of this understanding, in which so manifestly and yet so unexpressibly our whole life rests, may arise, by a sort of disengagedness, the mere looking-on of cognition, the perception of a thing of nature no longer bound to me, and in it, as an abstract momentum, sensation. Only by forgetting about the nourishing soil from which perception and cognition spring can the reality of the so-called external world become a problem. There are certain reasons why in the course of such abstraction and rarefaction intellect may come to exclaim: It is unthinkable that anything could be given but mere sensations, and thus challenged starts out to reconstruct the whole web out of this one thin thread. But what he weaves is a realistic theory rather than statements about what is given, and a theory at that, which is based on a bad foundation. One has not to prove, says Heidegger, that and how an external world exists, but to indicate why Dasein as being-in-the-world has a tendency first to bury the external world epistemologically into nothingness and then to prove it . . . After wrecking the original phenomenon of being-in-the-world one glues the remainder, the isolated subject, together with a world; but it sticks badly.

Being-in-the-world is primarily being-concerned-with, with such modes as manufacturing something, attending to, taking care of, employ, engage in, see through, question, consider, talk it over, determine, etc. Things are primarily implements on hand, hammer, table, bread. A chair is a chair to sit in, trees are lumber, wind is wind in the sails, persons are my wife or my dad, friend, colleague, the boy at the filling station, or "mein Führer." Only secondarily, by suspending my concern, the bare things of nature come into ken. Understanding of myself as being-with enables Dasein to understand other Dasein.

But this, like all understanding, is not knowledge gathered by cognition; it is rather a primary existential way of being which is a *conditio sine qua non* of cognition and knowledge.

Dasein is cast out into the world: Here I am; I know I am, but I know not whence and whither. On the other hand, Dasein projects itself into its own possibilities, running, feeling forward into the future, prefiguring, expecting, hoping, fearing, designing. A man at the wheel in heavy traffic is a good illustration. Thus *time* comes in by the anticipation of future possibilities, as future ever fading into past. I believe that this existential time is basic for our mathematical constructions, which on one side has its open field of possibilities generated by a process which may be iterated again and again, on the other side fixes the possibility by laws in order to obtain an individual number, an individual function, etc. With our mathematical construction, as I said some years before Heidegger's *Sein und Zeit* [*Being and Time*] appeared, we stand at that intersection of bondage and freedom which is the essence of man himself. Heidegger goes on to bind up existential time with guilt, conscience, and death. Here he pays his tribute to Kierkegaard, and I have the greatest difficulty in following him.

But enough, I think, is said to evoke the changed climate in which this philosophy grows as compared to idealism. The scientist may feel that clarity and precision have been sacrificed; he may even suspect a slackening of intellectual honesty. But in following the development of science we have seen that no other ground is left for science to build on than this dark but very solid rook which I once more call the concrete Dasein of man in his world. So the house stands on firm ground. But it has no roof, its construction reveals that the world is not closed but points beyond itself to what is represented by the symbols but can never be realized on the human level. The positivists try to nail a roof on by excluding the open sky which they do not see; that is blindness. Dogmatic religion and the theologians try to nail a roof on that pretends to cover the open sky together with the house on the rock; that is absurd. However, this is another story not to be told here and now.

I skipped what Heidegger has to say about discovery, awareness, speech, language, and truth. In living language man has tried to make his understanding explicit and communicable. I shall once more try to illustrate our basic view by an analogous situation in the realm of language. Out of our everyday language, which has often been denounced as a horsedealer's language, there has developed on the one hand the perfectly rational symbolism of mathematics, on the other hand poetry. In the living language both are contained, the poetic and the rational element. But people's living language is neither tarnished poetry nor a blurred substitute for mathematical symbolism; on the contrary, neither the one nor the other would and could exist without the nourishing stem of the language of our everyday life, with all it complexity, obscurity, crudeness, and ambiguity.[6]

I repeat once more that I feel far from sure myself about the daring things I have said, and probably I should have spoken in much more cautious words. Yet I honestly believe that a definite drift of thoughts in the direction which I have tried to describe makes itself felt, not only in mathematics and physics, but in our whole cultural life. Its interpretation is one of the tasks of the philosophers of our time.

8 ▣

The Unity of Knowledge

1954

The present solemn occasion on which I am given the honor to address you on our general theme "The Unity of Knowledge" reminds me, you will presently see why, of another Bicentennial Conference, held fourteen years ago by our neighborly university in the city of Brotherly Love. The words with which I started there a talk on "The Mathematical Way of Thinking" sound like an anticipation of today's topic; I repeat them: "By the mental process of thinking we try to ascertain truth; it is our mind's effort to bring about its own enlightenment by evidence. Hence, just as truth itself and the experience of evidence, it is something fairly uniform and universal in character. Appealing to the light in our innermost self, it is neither reducible to a set of mechanically applicable rules, nor is it divided into water-tight compartments like historical, philosophical, mathematical thinking, etc. True, nearer the surface there are certain techniques and differences; for instance, the procedures of fact-finding in a courtroom and in a physical laboratory are conspicuously different."[1] The same conviction was more forcibly expressed by the father of our Western philosophy, Descartes, who said: "The sciences taken all together are identical with human wisdom, which always remains one and the same, however applied to different subjects, and suffers no more differentiation proceeding from them than the light of the Sun experiences from the variety of the things it illumines."

But it is easier to state this thesis in general terms than to defend it in detail when one begins to survey the various branches of human knowledge. Ernst Cassirer, whose last years were so intimately connected with this university, set

194

out to dig for the root of unity in man by a method of his own, first developed in his great work *Philosophie der symbolische Formen* [*Philosophy of Symbolic Forms*]. The lucid "Essay on Man" written much later in this country and published by the Yale University Press in 1944, is a revised and condensed version. In it he tries to answer the question "What is man?" by a penetrating analysis of man's cultural activities and creations: language, myth, religion, art, history, science. As a common feature of all of them he finds: the symbol, symbolic representation. He sees in them "the threads which weave the symbolic net, the tangled net of human experience." "Man," he says, "no longer lives in a merely physical universe, he lives in a symbolic universe." Since "reason is a very inadequate term with which to comprehend the forms of man's cultural life in all their richness and variety," the definition of man as the *animal rationale* [rational animal] had better be replaced by defining him as an *animal symbolicum* [symbolic animal]. Investigation of these symbolic forms on the basis of appropriate structural categories should ultimately tend towards displaying them as "an organic whole tied together not by a *vinculum substantiale* [substantial bond], but a *vinculum functionale* [functional bond]." Cassirer invites us to look upon them "as so many variations on a common theme," and sets as the philosopher's task "to make this theme audible and understandable." Yet much as I admire Cassirer's analyses, which betray a mind of rare universality, culture, and intellectual experience, their sequence, as one follows them in his book, resembles more a suite of bourrées, sarabands, menuets, and gigues than variations on a single theme. In the concluding paragraph he himself emphasizes "the tensions and frictions, the strong contrasts and deep conflicts between the various powers of man, that cannot be reduced to a common denominator." He then finds consolation in the thought that "this multiplicity and disparateness does not denote discord or disharmony," and his last word is that of Heraclitus: "Harmony in contrariety, as in the case of the bow and the lyre." Maybe man cannot hope to be more than that; but am I wrong when I feel that Cassirer quits with a promise unfulfilled?[2]

In this dilemma let me now first take cover behind the shield of that special knowledge in which I have experience through my own research: the natural sciences including mathematics. Even here doubts about their methodical unity have been raised. This, however, seems unjustified to me. Following Galileo, one may describe the method of science in general terms as a combination of passive observation refined by active experiment with that symbolic construction to which theories ultimately reduce. Physics is the paragon. Hans

Driesch and the holistic school have claimed for biology a methodical approach different from, and transcending, that of physics. However, nobody doubts that the laws of physics hold for the body of an animal or myself as well as for a stone. Driesch's attempts to prove that the organic processes are incapable of mechanical explanation rest on a much too narrow notion of mechanical or physical explanation of nature. Here quantum physics has opened up new possibilities. On the other side, wholeness is not a feature limited to the organic world. Every atom is already a whole of quite definite structure; its organization is the foundation of possible organizations and structures of the utmost complexity. I do not suggest that we are safe against surprises in the future development of science. Not so long ago we had a pretty startling one in the transition from classical to quantum physics. Similar future breaks may greatly affect the epistemological interpretation, as this one did with the notion of causality; but there are no signs that the basic method itself, symbolic construction combined with experience, will change.

It is to be admitted that on the way to their goal of symbolic construction scientific theories pass preliminary stages, in particular the classifying or morphological stage. Linnaeus' classification of plants, Cuvier's comparative anatomy are early examples; comparative linguistics or jurisprudence are analogues in the historical sciences. The features which natural science determines by experiments, repeatable at any place and any time, are *universal*; they have that empirical necessity which is possessed by the laws of nature. But beside this domain of the necessary there remains a domain of the *contingent*. The one cosmos of stars and diffuse matter, Sun and Earth, the plants and animals living on earth, are accidental or singular phenomena. We are interested in their evolution. Primitive thinking even puts the question "How did it come about?" before the question "How is it?" All history in the proper sense is concerned with the development of one singular phenomenon: human civilization on earth. Yet if the experience of natural science accumulated in her own history has taught one thing, it is this, that in its field knowledge of the laws and of the inner constitution of things must be far advanced before one may hope to understand or hypothetically reconstruct their genesis. For want of such knowledge as is now slowly gathered by genetics, the speculations on pedigrees and phylogeny let loose by Darwinism in the last decades of the nineteenth century were mostly premature. Kant and Laplace had the firm basis of Newton's gravitational law when they advanced their hypotheses about the origin of the planetary system.

After this brief glance at the methods of natural science, which are the same in all its branches, it is time now to point out the limits of science. The riddle posed by the double nature of the ego certainly lies beyond those limits. On the one hand, I am a real individual man; born by a mother and destined to die, carrying out real physical and psychical acts, one among many (far too many, I may think, if boarding a subway during rush hour). On the other hand, I am "vision" open to reason, a self-penetrating light, immanent sense-giving consciousness, or however you may call it, and as such unique. Therefore I can say to myself both: "I think, I am real and conditioned" as well as "I think, and in my thinking I am free." More clearly than in the acts of volition the decisive point in the problem of freedom comes out, as Descartes remarked, in the theoretical acts. Take for instance the statement $2 + 2 = 4$: not by blind natural causality, but because I *see* that $2 + 2 = 4$ does this judgement as a real psychic act form itself in me, and do my lips form these words: two and two make four. Reality or the realm of Being is not closed, but open toward Meaning in the ego, where Meaning and Being are merged in indissoluble union—though science will never tell us how. We do not see through the real origin of freedom.

And yet, nothing is more familiar and disclosed to me than this mysterious "marriage of light and darkness," of self-transparent consciousness and real being that I am myself. The access is my knowledge of myself from within, by which I am aware of my own acts of perception, thought, volition, feeling, and doing, in a manner entirely different from the theoretical knowledge that represents the "parallel" cerebral processes in symbols. This inner awareness of myself is the basis for the more or less intimate understanding of my fellow men, whom I acknowledge as beings of my own kind. Granted that I do not know of their consciousness in the same manner as of my own, nevertheless my "interpretative" understanding of it is apprehension of indisputable adequacy. As hermeneutic interpretation it is as characteristic for the historical, as symbolic construction is for the natural sciences. Its illumining light not only falls on my fellow-men; it also reaches, though with ever increasing dimness and incertitude, deep into the animal kingdom. Kant's narrow opinion that we can feel compassion, but cannot share joy with other living creatures, is justly ridiculed by Albert Schweitzer who asks: "Did he never see a thirsty ox coming home from the fields, drink?" It is idle to disparage this hold on nature "from within" as anthropomorphic and elevate the objectivity of theoretical construction, though one must admit that understanding, for the very reason that it is *concrete* and *full*, lacks the freedom of the "hollow symbol." Both roads run, as it were, in opposite directions: what is darkest for theory, man, is the most luminous for the understanding from within; and to the elementary inorganic processes, that are most easily approachable by theory, interpretation finds no access whatever. In biology the latter may serve as a guide to important problems, although it will not provide an objective theory as their solution. Such teleological statements as "The hand is there to grasp, the eye to see" drive us to find out what internal material organization enables hand and eye, according to the physical laws (that hold for them as for any inanimate object), to perform these tasks.

I will not succumb to the temptation of foisting Professor Bohr's idea of complementarity upon the two opposite modes of approach we are discussing here. However, before progressing further, I feel the need to say a little more about the constructive procedures of mathematics and physics.

Democritus, realizing that the sensuous qualities are but effects of external agents on our sense organs and hence mere apparitions, said: "Sweet and bitter, cold and warm, as well as the colors, all these exist but in opinion and not in reality; what really exist are unchangeable particles, atoms, which move in empty space." Following his lead, the founders of modern science, Kepler, Galileo, Newton, Huygens, with the approval of the philosophers, Descartes, Hobbes, Locke, discarded the sense qualities, on account of their subjectivity, as building material of the objective world which our perceptions reflect. But they clung to the objectivity of space, time, matter, and hence of motion and the corresponding geometric and kinematic concepts. Thus Huygens, for instance, who developed the undulatory theory of light, can say with the best of conscience that colored light beams are *in reality* oscillations of an ether consisting of tiny particles. But soon the objectivity of space and time also became suspect. Today we find it hard to realize why their intuition was thought particularly trustworthy. Fortunately Descartes' analytic geometry had provided the tool to get rid of them and to replace them by numbers, i.e., mere symbols. At the same time one learned how to introduce such concealed characters, as, e.g., the inertial mass of a body, not by defining them explicitly, but by postulating certain simple laws to which one subjects the observation of reacting bodies. The upshot of it all is a purely symbolic construction that uses as its material nothing but mind's free creations: symbols. The monochromatic beam of light, which for Huygens was in reality an ether wave, has now become a formula expressing a certain undefined symbol F, called electromagnetic field, as a mathematically defined function of four other symbols x, y, z, t, called space-time coordinates. It is evident that now the words "in reality" must be put between quotation marks; who could seriously pretend that the symbolic construct is the true real world? Objective Being, reality, becomes elusive; and science no longer claims to erect a sublime, truly objective world above the Slough of Despond in which our daily life moves. Of course, in some way one must establish the connection between the symbols and our perceptions. Here, on the one hand, the symbolically expressed laws of nature (rather than any explicit "intuitive" definitions of the significance of the symbols) play a fundamental role, on the other hand the concretely described procedures of observation and measurement.

In this manner a theory of nature emerges which only as a whole can be confronted with experience, while the individual laws of which it consists, when taken in isolation, have no verifiable content. This discords with the traditional idea of truth, which looks at the relation between Being and Knowing from the side of Being, and may perhaps be formulated as follows: "A statement points to a fact, and it is true if the fact to which it points is so as it states." The truth of physical theory is of a different brand.

Quantum theory has gone even a step further. It has shown that observation always amounts to an uncontrollable intervention, since measurement of one quantity irretrievably destroys the possibility of measuring certain other quantities. Thereby the objective Being which we hoped to construct as one big piece of cloth each time tears off; what is left in our hands are—rags.

The notorious man-in-the-street with his common sense will undoubtedly feel a little dizzy when he sees what thus becomes of that reality which seems to surround him in such firm, reliable and unquestionable shape in his daily life. But we must point out to him that the constructions of physics are only a natural prolongation of operations his own mind performs (though mainly unconsciously) in perception, when, e.g., the solid shape of a body constitutes itself as the common source of its various perspective views. These views are conceived as appearances, for a subject with its continuum of possible positions, of an entity on the next higher level of objectivity: the three-dimensional body. Carry on this "constitutive" process in which one rises from level to level, and one will land at the symbolic constructs of physics. Moreover, the whole edifice rests on a foundation which makes it binding for all reasonable thinking: of our complete experience it uses only that which is unmistakably *aufweisbar* [provable, demonstrable].

Excuse me for using here the German word. I explain it by reference to the foundations of mathematics. We have come to realize that isolated statements of classical mathematics in most cases make as little sense as do the statements of physics. Thus it has become necessary to change mathematics from a system of meaningful propositions into a game of formulas which is played according to certain rules. The formulas are composed of certain clearly distinguishable symbols, as concrete as the men on a chess board. Intuitive reasoning is required and used merely for establishing the consistency of the game—a task which so far has only partially been accomplished and which we may never succeed in finishing. The visible tokens employed as symbols must be, to repeat Hilbert's words, "recognizable with certainty, independently of time and place, and independently of minor differences and the material conditions of their execution (e.g., whether written by pencil on paper or by chalk on blackboard)." It is also essential that they should be reproducible where- and whenever needed. Now here is the prototype of what we consider as *aufweisbar*, as something to which we can point *in concreto*. The inexactitude which is inseparable from continuity and thus clings inevitably to any spatial configurations is overcome here in principle, since only clearly distinguishable marks are used and slight modifications are ignored "as not affecting their identity." (Of course, even so errors are not excluded.) When putting such symbols one behind the other in a formula, like letters in a printed word, one obviously employs space and spatial intuition in a way quite different from a procedure that makes space in the sense of Euclidean geometry with its exact straight lines, etc., one of the bases on which knowledge rests, as Kant does. The *Aufweisbare* we start with is not such a pure distillate, it is much more concrete.

Also the physicist's measurements, e.g., reading of a pointer, are operations performed in the *Aufweisbaren*—although here one has to take the approximate character of all measurements into account. Physical theory sets the mathematical formulas consisting of symbols into relation with the results of concrete measurements.

At this juncture I wish to mention a collection of essays by the mathematician and philosopher Kurt Reidemeister published by Springer in 1953 and 1954 under the titles *Geist und Wirklichkeit* [*Spirit and Reality*] and *Die Unsachlichkeit der Existenzphilosophie* [*The Subjectivity of Existential Philosophy*]. The most important is the essay "Prolegomena einer kritischen Philosophie" ["Prolegomena to a Critical Philosophy"] in the first volume. Reidemeister is positivist in as much as he maintains the irremissible nature of the factual which science determines; he ridicules (rightly, I think) such profound sounding but hollow evocations as Heidegger indulges in, especially in his last publications. On the other hand, by his insistence that science does not make use of our full experience, but selects from it that which is *aufweisbar*, Reidemeister makes room for such other types of experience as are claimed by the windbags of profundity as their proper territory: the experience of the indisposable significant in contrast to the disposable factual. Here belongs the intuition through and in which the beautiful, whether incorporated in a vase, a piece of music or a poem, appears and becomes transparent, and the reasonable experience governing our dealings and communications with other people; an instance: the ease with which we recognize and answer a smile. Of course, the physical and the aesthetic properties of a sculpture are related to each other; it is not in vain that the sculptor is so exacting with respect to the geometric properties of his work, because the desired aesthetic effect depends on them. The same connection is perhaps even more obvious in the acoustic field. Reidemeister, however, urges us to admit our *Nicht-Wissen*, our not knowing how to combine these two sides by theory into one unified realm of Being—just as we cannot see through the union of I, the conditioned individual, and I who thinking am free. This *Nicht-Wissen* is the protecting wall behind which he wants to save the indisposable significant from the grasp of hollow profundity and restore our inner freedom for a genuine apprehension of ideas. Maybe, I overrate Reidemeister's attempt, which no doubt is still in a pretty sketchy state, when I say that, just as Kant's philosophy was based on, and made to fit, Newton's physics, so his attempt takes the present status of the foundations of mathematics as its lead. And as Kant supplements his *Critique of Pure Reason* by one of practical reason and of aesthetic judgement, so leaves Reidemeister's analysis room for other experiences than science makes use of, in particular for the hermeneutic understanding and interpretation on which history is based.

Let me for the few remarks I still want to make adopt the brief terms science and history for natural and historical sciences (*Natur- und Geistes-Wissenschaften*). The first philosopher who fully realized the significance of hermeneutics as the basic method of history was Wilhelm Dilthey. He traced it back to the exegesis of the Holy Script. The chapter on history in Cassirer's "Essay on Man" is one of the most successful. He rejects the assumption of a special historical logic or reason as advanced by Windelband or more recently and much more impetuously by Ortega y Gasset. According to him the essential difference between history and such branches of science as, e.g., palaeontology dealing with singular phenomena lies in the necessity for the historian to interpret his "petrefacts," his monuments and documents, as having a symbolic content.

Summarizing our discussion I come to this conclusion. At the basis of all knowledge there lies: (1) *Intuition*, mind's originary act of "seeing" what is given to him; limited in science to the *Aufweisbare*, but in fact extending far beyond these boundaries. How far one should go in including here the *Wesensschau* of Husserl's phenomenology, I prefer to leave in the dark.[3] (2) *Understanding and expression*. Even in Hilbert's formalized mathematics I must understand the directions given me by communication in words for how to handle the symbols and formulas. Expression is the active counterpart of passive understanding. (3) *Thinking the possible*. In science a very stringent form of it is exercised when, by thinking out the possibilities of the mathematical game, we try to make sure that the game will never lead to a contradiction; a much freer form is the imagination by which theories are conceived. Here, of course, lies a source of subjectivity for the direction in which science develops. As Einstein once admitted, there is no logical way leading from experience to theory, and yet the decision as to which theories are adopted turns out ultimately to be unambiguous. Imagination of the possible is of equal importance for the historian who tries to re-enliven the past. (4) On the basis of intuition, understanding and thinking of the possible, we have in science: certain practical actions, namely the *construction* of symbols and formulas on the mathematical side, the construction of the measuring devices on the empirical side. There is no analogue for this in history. Here its place is taken by *hermeneutic interpretation*, which ultimately springs from the inner awareness and knowledge of myself. Therefore the work of a great historian depends on the richness and depth of his own inner experience. Cassirer finds wonderful words for Ranke's intellectual and imaginative, not emotional, sympathy, the universality of which enabled him to write the history of the Popes and of the Reformation, of the Ottomans and the Spanish Monarchy.

Being and Knowing, where should we look for unity? I tried to make clear that the shield of Being is broken beyond repair. We need not shed too many tears about it. Even the world of our daily life is not *one*, to the extent people are inclined to assume; it would not be difficult to show up some of its cracks. Only on the side of Knowing there may be unity. Indeed, mind in the fullness of its experience has unity. Who says "I" points to it. But just because it is unity,

I am unable to represent it otherwise than by such characteristic actions of the mind mutually supporting each other as I just finished enumerating. Here, I feel, I am closer to the unity of the luminous center than where Cassirer hoped to catch it: in the complex symbolic creations which this lumen built up in the history of mankind. For these, and in particular myth, religion, and alas!, also philosophy, are rather turbid filters for the light of truth, by virtue, or should I say, by vice of man's infinite capacity for self-deception.[4]

What else than turbidity could you then have expected from a philosophical talk like this? If you found it particularly aimless, please let me make a confession before asking for your pardon. The reading of Reidemeister's essays has caused me to think over the old epistemological problems with which my own writings had dealt in the past; and I have not yet won through to a new clarity. Indecision of mind does not make for coherence in speaking one's mind. But then, would one not cease to be a philosopher, if one ceased to live in a state of wonder and mental suspense?

9 ▣

Insight and Reflection

1955

I should like to take this opportunity of describing the part which philosoph-ical reflection, along with scientific insight, has played in my life. Although my work has centered on mathematical research, with occasional detours into theoretical physics, I was always feeling the urge to render a reflective account of the meaning and goal of that research. In a lecture on "The Levels of Infin-ity," following upon a discussion of constructive mathematics and reflective metamathematics, I once described their mutual relation in this fashion:

> In the intellectual life of man we find discernibly separated, on the one hand, a sphere of *action*, of shaping and constructing to which the active artist, scientist, engineer, and statesman are dedicated and which is governed in the field of science by the norm of objectiv-ity; and on the other hand, the sphere of *reflection*, which fulfills itself in insights and judgments and which, as the struggle of gain-ing insight into the meaning of our actions, is to be considered the proper domain of the philosopher. The danger faced by the work of creation, if not controlled by reflection, is that it outruns reason, goes astray, and hardens into routine; the danger of reflection is that it becomes just noncommittal "talk about it," paralyzing man's creative powers.[1]

As I set about recounting my life at least insofar as philosophical urges played a part in it, it will be inevitable that I shall have to touch on the great themes of space and time, the material world, man and I, and God, in this order.

I have never forgotten how, during my next to last year in school, I happened in the attic of my parent's home to come upon a copy of a short commentary on Kant's *Critique of Pure Reason*, dating from 1790, yellowed and foxed by long storage. Kant's teaching on the "ideality of space and time" immediately took powerful hold of me; with a jolt I was awakened from my "dogmatic slumber," and the mind of the boy found the world being questioned in radical fashion.

Is it necessary to repeat the quintessence of Kant's teachings here? He recognized that time and space are not inherent in the objects of a world, existing as such and independently of our awareness, but rather that they are *conceptual forms* which are based in our intellects. As such, he placed them apart from the hylic foundation of perceiving and sensing.[2] I quote: "Since that within which alone the sensory perceptions can be ordered and can be combined into certain forms, cannot be in turn itself a sensory perception, it is therefore true that, although the material aspects of all appearances are given to us only a posteriori, the forms of the latter must all be present in the mind a priori, and thus capable of being contemplated independently of all sensory perception." Or, as Fichte says in his strong and always a bit eccentric language: "Transparent, penetrable space, the purest image of my knowing, cannot be inspected but must be seen intuitively, and within it my inspecting itself is so seen. The light is not outside of me, but rather in me."

This teaching seemed at a stroke to explain an almost universally accepted fact, namely, the circumstance that the basic principles of geometry appear to us immediately and compellingly obvious, without our having to take recourse to experience. Kant differentiates *analytical* judgments, which do nothing more than to state what is contained already in the concepts, such as "A round thing is round," or "If Socrates is a man and all men are mortal, then Socrates is mortal," from *synthetic* judgments, of which Newton's law of gravity provides an example. That analytical judgments are valid a priori, independently of experience, is no surprise. Yet the statements of geometry provide, in view of the above, an example of synthetic judgments which, despite their synthetic character, are valid a priori, neither based on experience nor permitting their indispensability to be shaken by any experience. Kant's central question was *"How can there be synthetic judgments a priori?"* and his insight into the nature of space furnished an answer, insofar as the theorems of geometry are at issue.

While I had no trouble in making this part of Kant's teachings my own, I still had much trouble with the *Schematism of Pure Mental Concepts* when

I entered the University of Göttingen in 1904. David Hilbert was teaching there, having recently published his epoch-making book *The Foundations of Geometry*. From its pages breathed the spirit of the modern axiomatic approach. In a completeness which leaves Euclid far behind, it establishes the axioms of geometry. Moreover, it examines their logical interdependence not only by drawing on the so-called non-Euclidean geometry, then nearly a century old, but by constructing, mostly on an arithmetical basis, a plethora of other strange geometries. Kant's bondage to Euclidean geometry now appeared naive. Under this overwhelming blow, the structure of Kantian philosophy, on which I had hung with faithful heart, crumbled into ruins.

Here I interrupt my story in order to indicate briefly what today appears to me as the reasonable attitude on the problem of space. Firstly, the special theory of relativity has fused space-time in the cosmos into a single four-dimensional continuum. Secondly, it has turned out to be essential to distinguish the amorphous continuum which is treated today in the so-called discipline of topology, from its structure, in particular its metric structure. Physical geometry, founded on a physically verifiable concept of congruence, was already viewed by Newton as a part of mechanics based on experience. He says: "Geometry has its foundation in mechanical practice and is in effect nothing more than the particular part of the whole of mechanics which puts the art of measurement on precise and firm foundations." Helmholtz showed that the two parts of Kant's doctrine—(1) space is a pure form of intuition, and (2) the science of space, Euclidean geometry, is valid a priori—are not so closely interconnected that part (2) would follow from (1). He is prepared to accept (1) as a correct expression of the state of affairs, but he argues that nothing more can be concluded from it than that all things in the external world have spatial extent. In agreement with Newton and Riemann, he then develops the empirical-physical meaning of geometry.[3]

Riemann's comment that "the empirical concepts on which the determination of measures is founded, the concept of the rigid body and of the light ray, lose their meaning on the level of the infinitely small," has later become a source of worry for the quantum physicists.[4] In the meantime, however, the infinitesimal geometry of manifolds of arbitrary dimension, as established by Riemann, came into its own in Einstein's general theory of relativity. Going beyond Riemann, it showed that the metric structure prevailing in the real four-dimensional world is not an entity given in advance, but is influenced by physical processes and influences them in turn: It is in the phenomena of gravitation where the fluidity of the metric field manifests itself.

If, besides the physical space, one recognizes an intuitive one and claims that, for reasons of its nature, its structure must fulfill the Euclidean laws, this does not necessarily contradict our physical insight, inasmuch as the latter also holds to the Euclidean structure—bluntly speaking, to the validity of Pythagoras's theorem, in any infinitely small neighborhood of a point O (in which the self is momentarily located). However, one must then concede that the relationship between the intuitive and the physical space becomes vaguer as one recedes further from the self-center O. The former can be compared to a tangent plane which touches a curved surface, i.e., the physical space, at point O. In the immediate neighborhood of O they coincide, but the further one moves from O the more arbitrary becomes the continuation of this coincidence relation as a one-to-one correspondence between surface and plane.

In the physical world, as I have said, time has been fused with space into a unified, four-dimensional continuum. In confirmation of Leibniz's thesis that the division of past and future rests in the causal structure of the four-dimensional world, the theory of relativity led to a description of this structure in which, deviating from traditional ideas, the simultaneity as well as the spatial coincidence of events lose their objective meaning. In this world, my body, if I consider it as a point, traces out a one-dimensional world-line, along which it is possible physically to define a proper time. On this line, the order represented by the words "past," "present," "future" of course exists. The temporal order of phenomena, inherent as their general form in the awareness acts of the I, cannot be placed in parallel with the time coordinate of the four-dimensional world continuum, but has its physical counterpart in just this proper time along the world-line of the I-body (fig. 9.1).

Moreover, it also turns out that within the framework of the general theory of relativity it is possible, in a certain objective sense which is not based on the Kantian distinction between source and mode of cognition, to separate a priori from a posteriori aspects of physical space. The confrontation is between the unique, absolutely given Euclidean-Pythagorean nature of the metric, which does not share the unavoidable vagueness of whatever occupies a variable position on a continuous scale, and the relative orientation of the metrics at different points, i.e., the arbitrary quantitative course of the metric

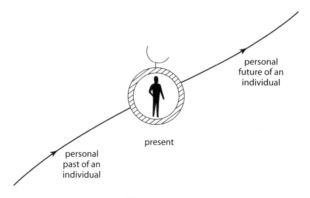

Figure 9.1

field, dependent on the distribution of matter, ever-changing, and capable of determination only by direct recourse to empirical reality.

At one time I undertook to explain exactly on the basis of this separation the specifically Pythagorean nature of the metric in its mathematical uniqueness.[5] One faces a similar problem when one attempts to understand why the world happens to be four-dimensional, rather than having some other number of dimensions. One must bear in mind that all the physical laws which we know so far (and the geometric ones which are relevant for them) can be transferred in a fully consistent manner to an arbitrary number of dimensions. Hence, there is nothing in them which would in any way single out the dimension four. Mathematics, however, informs us of entities, especially in group theory, whose structure varies greatly depending on their dimension. Apparently, physics with its present laws has not yet pushed to the depths where it would need this type of mathematics. That is why we cannot at the present give a really convincing answer to the question of the basis for the world's four-dimensionality; and I will also let it be a moot point whether my attempt to explain the Pythagorean nature of the metric field has hit the mark.

Enough of space and time. I now continue my tale. After contact with modern mathematics had destroyed my belief in Kant, I turned with fervor to the study of mathematics. Whatever remained of epistemological interests was satisfied by such writings as *Science and Hypothesis* by Henri Poincaré, the works of Ernst Mach, and the well-known *History of Materialism* by Friedrich Albert Lange.

The next epochal event for me was that I made an important mathematical discovery. It concerned the regularity in the distribution of the eigenfrequencies of a continuous medium, like a membrane, an elastic body, or the electromagnetic ether. The idea was one of many, as they probably come to every young person preoccupied with science, but while the others soon burst like soap bubbles, this one led, as a short inspection showed, to the goal. I was myself rather taken aback by it, as I had not believed myself capable of anything like it. Added to it was the fact that the result, although conjectured by the physicists some time ago, appeared to most mathematicians as something whose proof was still in the far future. While I was feverishly working on the proof, my kerosene lamp had begun to smoke, and I was no sooner finished than thick sooty flakes began to rain down from the ceiling onto my paper, hands, and face.

Gottfried Keller was uninhibited enough to confess that it was his love for a woman which shook his belief in immortality. She was Johanna Kepp, whose views had been influenced by her father's close friend, the materialist philosopher Ludwig Feuerbach. A similar thing happened to me: My peace of mind in positivism was shaken when I fell in love with a young singer whose life was grounded in religion and who belonged to a circle that was led philosophically by a well-known Hegelian. Partly because of my immaturity, but also because of this unbridgeable chasm in philosophical outlook, nothing came of it. The shock, however, continued to work. It was not long afterward that I married a philosophy student, a disciple of Edmund Husserl, the founder of phenomenology who was then working in Göttingen.

So it came to be Husserl who led me out of positivism once more to a freer outlook upon the world. At the same time, I had to complete the change from an unsalaried university lecturer's post at Göttingen to the responsibilities of a professor of geometry at the Federal Institute of Technology in Zurich. There, through the assistance of Medicus, whose seminar my wife visited, she and I were led to Fichte's philosophy of science.[6] Metaphysical idealism, toward which Husserl's phenomenology was then shyly groping, here received its most candid and strongest expression. It captured my imagination, even though I had to concede to my wife, who was more at home with Husserl's careful methodology than with Fichte's dash, that Fichte could not help being swept to ever more abstruse constructions by his stubbornness, which made him blind to facts and reality in the pursuit of an idea.

Husserl's work had its origins in mathematics. In the pursuit of his *Logical Investigations*, and in part under the influence of the philosopher Franz

Brentano, he became the adversary of the psychologism which prevailed at the turn of the century. Instead, he developed the method of phenomenology, whose goal it was to capture the phenomena in their essential being—purely as they yield themselves apart from all genetical and other theories in their encounter with our consciousness. This quintessential examination unfolded to him a far broader field of evidently valid a priori insights than the twelve principles which Kant had posited as the constituting foundation of the world of experience. I quote from his systematic exposition of 1922, *Ideas for a Pure Phenomenology and a Phenomenological Philosophy*:

"The *What* of things, 'cast in idea,' is being. The comprehension of pure being through awareness thereof does not imply the slightest postulation of any one specific individual existence. Pure phenomenological truths do not make the slightest assertion about facts; from them alone, therefore, not even the most minuscule factual truth can be concluded."

On another page, however, Husserl states: "Every description of being which can be related to modes of experience, expresses an unconditionally valid norm for possible empirical existence." Typical of the phenomenological method is this affirmation: "The immediate 'seeing,' not just the sensory, experiencing sight but seeing in general, as given in originary consciousness of whatever kind, is the ultimate source of justification for all reasonable assertions. What offers itself to us as originating in our intuition, is simply to be accepted in the form in which it gives itself, but only within the limits within which it gives itself."

To point up the antithesis between an accidental, factual law of nature and a necessary law of being, Husserl cites the following two statements: "All bodies are heavy" and "All bodies have spatial extent." Perhaps he is right, but one senses even in this first example how uncertain generally stated epistemological distinctions become as soon as one descends from generality to specific concrete applications. In my book *Space-Time-Matter*, which first appeared in 1918 and in which I published my lectures on the general theory of relativity, I commented on this as follows:

> The inquiries which we have conducted here about space seem to me a good example for the analysis of being that is aspired to by phenomenological philosophy, an example that is typical of such cases where non-immanent entities are concerned. We see in the historical development of the problem of space how difficult it is for us who, as

human beings, are enmeshed in reality to hit the crucial point. A long mathematical evolution, the great revelations of the geometric studies from Euclid to Riemann, the penetration of nature and her laws by physics since the time of Galileo, with ever new boosts from empirical observation, and finally the genius of a few singular minds—Newton, Gauss, Riemann, Einstein—all were necessary to tear us loose from the accidental, non-essential characteristics to which we are at first so attached. Once the new, more comprehensive viewpoint has won out, however, reason sees the light; she recognizes and acknowledges what she has thus understood "out of herself." On the other hand, she never had the strength (although she was "on the scene" throughout the entire development of the problem) to see it through at a single stroke. This must be held up to the impatience of the philosophers who feel that, on the basis of a single act of exemplary concentration, they are able to give an adequate description of being. The example of space is at the same time most instructive with regard to the particular question of phenomenology that appears to me the decisive one: To what extent does the limitation of those aspects of being which are finally revealed to consciousness express an innate structure of what is given, and to what extent is this a mere matter of convention?

In general, this is the view of the relation between cognition and reflection which I still hold today. Einstein's development of the general theory of relativity, and of the law of gravity which holds true in the theory's framework, is a most striking confirmation of this method which combines experience based on experiments, philosophical analysis of existence, and mathematical construction. Reflection on the meaning of the concept of motion was important for Einstein, but only in such a combination did it prove fruitful.

Yet the main issue under consideration in Husserl's great work of 1922 is the relationship between the immanent consciousness, the pure ego from whence all its actions emanate, and the real psychophysical world, upon whose objects these acts are intentionally directed. The term "intentional" was borrowed by Franz Brentano from scholastic philosophy, and Husserl appropriated it. Even the experiences of awareness themselves can, in reflection, become the intentional object of immanent perceptions that are directed at them. The intentional object of an external perception—yonder tree, for example— is the thing as it gives itself in the perception, without the question being raised of whether and in what sense it conforms with a thus or similarly constituted real tree.

Husserl laboriously describes the phenomenological *epoché*, through which the general thesis of the world's real existence, as the essence of the natural attitude about the world, is put out of action—"put in parentheses," as it were.[7] "Consciousness," he says, "has an innate existence in itself, whose absolute self-being is not affected by this elimination; thus there remains a pure consciousness—as a phenomenological residuum." Concerning space as an object, Husserl says that, with all its transcendence, it is something that is perceived and given in material irrefutability to our awareness. Sensory data, "shaded off" in various ways within the concrete unity of this perception and enlivened by comprehension, fulfill in this manner their representative "function"; in other words, they constitute in unison with this quickened comprehension what we recognize as "appearances of" color, form, etc.

I do not find it so easy to agree with this. At any rate, one cannot disavow that the particular manner in which, through this function of inspiration, an identifiable object is placed before me, is guided by a great number of earlier experiences—no matter how much one may rebel against Helmholtz's turn of phrase concerning "unconscious" conclusions. The theoretical-symbolic construction, through which physics attempts to comprehend the transcendental content that lies behind the observations, is far from inclined to stop with this corporeally manifested identity. I should, therefore, say that Husserl describes but one of the levels which has to be passed in the endeavor through which the external world is constituted. In the consciousness he distinguishes between a hylic and a noetic layer and between the sensual *hylē* and the intentional *morphē*.[8] Thus he speaks of the manner in which (for example, as regards nature) noeses breath spirit into matter and weave themselves into a fabric of manifold-uniform continua and syntheses, thus bringing about awareness of something in such a fashion that an objective unity of substantive identity can therein be concordantly certified, *demonstrated*, and *reasonably* ascertained. He continues emphatically: "Awareness is awareness of something; it is its nature to be the sense, so to speak, the quintessence of Soul, Mind, and Reason. It is not a title for 'psychic complexes,' for commingled contents, for 'bundles' or currents of perception which, being meaningless, would also fail, no matter in what mixture, to yield meaning; it is rather consciousness through and through, the source of all reason and unreason, of all justice and injustice, of all reality and fiction, of all that is worthy and unworthy, of all deeds and misdeeds."

Concerning the antithesis of experience and object, Husserl claims no more than merely phenomenal existence for the transcendental as it is given in its various shadings, in opposition to the absolute existence of the immanent; i.e., the certitude of the immanent in contrast to the uncertainty of the transcendental perception. The thesis of the world in its accidental arbitrariness thus stands face to face with the thesis of the pure I and the I-life which is indispensable and, for better or worse, unquestionable. "Between awareness and reality there yawns a veritable chasm of meaning," he says. "Immanent existence has the meaning of absolute being which *'nulla re indiget ad existendum'* [no thing remains in existence]; on the other hand the world of the transcendental *'res'* [thing] is completely dependent on awareness,—dependent, moreover, not just on being logically thinkable but on actual awareness."

Here finally arises in its full seriousness the metaphysical question concerning the relation between the one pure I of immanent consciousness and the particular lost human being which I find myself to be in a world full of people like me (for example, during the afternoon rush hour on Fifth Avenue in New York). Husserl does not say much more about it than that "only through experience of the relationship to the body does awareness take on psychological reality in man or animal." Yet he immediately insists again on the autonomous character of pure consciousness, undiminished in its essential nature by such interweaving with perceptions, i.e., by these psychophysiological referrals to the corporeal. "All real entities are entities of the intellect. Intellectual entities presuppose the existence of a consciousness which assigns them their meaning and which, in its turn, exists absolutely and not as the result of assigned meaning." It should, therefore, be clear "that, despite all the talk, to be sure most sensible and well-reasoned, about the real existence of the human ego and the experiences of its consciousness within the world, any reflection about pure consciousness has to accept it as a connected form of being which is closed in itself, into which nothing can enter and from which nothing can escape, which cannot be causally affected nor itself causally affect anything. The entire spatio-temporal world, on the other hand, into which man and human self consider themselves ranged as subordinate specific realities, has by its very nature only intentional existence,—in other words, an existence in but the secondary relative sense of existing for a consciousness. It is an existence which consciousness establishes in its experiences, which—as a matter of principle—can only be beheld and defined as what is identical about a concordantly motivated manifold of appearances,—*but which is nothing beyond this.*"

Fichte states the basic position of epistemological idealism even more radically than Husserl. He is everything but a phenomenologist; he is a constructivist of the purest form who, without looking left or right, follows his stubborn path of construction. He reminds me in many ways of Saint Paul: Both have the same—as it were—uncouth way of thinking which ends up by sweeping you along because of its solid precision. They show the same complete indifference toward experience—in the case of Paul, especially toward the testimonials of Christ's actual life. They have the same stiff-necked faith, intolerant of all contradiction, in an extravagant construction which shows itself in Fichte, for instance, when he uses such phrases as: "It must be so, and it cannot be otherwise; therefore thus it is" or in the title of a pamphlet: "A Report Clear as the Sun to the General Public about the Intimate Nature of the Newest Philosophy; An Attempt to Coerce the Reader's Understanding." Common to both of them as zealots is their occasionally uncontrolled vituperation against those who believe differently. Confronting dogmatism with idealism as the only two possible philosophies, Fichte makes a remark which sounds like an anticipation of existentialism: "The kind of philosophy one chooses depends upon what kind of human being he is"; but he follows it up at once with the zealot's comment: "A character, weak by nature or made so by spiritual servility, by sophisticated luxury and vanity, will never be able to raise itself to Idealism."

Fichte describes his method as follows:

> The summons is issued to think a certain concept or factual context. The inevitable manner in which to perform that act is founded in the nature of the intelligence and, in contrast to the specific act of thinking itself, does not depend upon anything arbitrary. This manner is something that is *inevitable*, but which occurs only in a free action,—something that is discovered but the discovery of which is contingent on freedom. To this extent, idealism demonstrates in the immediate consciousness what it claims to be true. It is, however, mere conjecture that this inevitability is the basic law of all of reason, from which one can derive the whole system of our necessary ideas,—not only those which concern a universe and the determination of its objects by the power of ordering and reflecting judgment, but also those which concern ourselves, as free and practical beings under laws.[9]

He can only prove this conjecture by the actual derivation in which he demonstrates that what was established at the start as the basic law, and shown immediately to be present in the consciousness, cannot be possible unless at the same time something else happens, and that the latter is not possible without a third thing happening at the same time, and so on. The system of the ideas, so derived, is equated to the sum total of experience; they find their confirmation in experience; the a priori therefore coincides in the end with the a posteriori. This sounds as if the world was to be deduced, not only in terms of the *possibilities* that are rooted in its structure, but also in its unique *factuality*. The actual execution of this program by Fichte I can only describe as preposterous. In the antithesis of constructivism and phenomenology, my sympathies lie entirely on his side; yet how a constructive procedure which finally leads to the symbolic representation of the world, not a priori, but rather with continual reference to experience, can really be carried out, is best shown by physics—above all in its two most advanced stages: the theory of relativity and quantum mechanics.

Speaking of the I, Fichte says: "The I demands, that it comprise all reality and fill up infinity. This demand is based, as a matter of necessity, on the idea of the infinite I, simply posited by itself; this is the absolute I (which is not the I given in real awareness). The I has to reflect on itself; that likewise lies in its meaning." Referring to the I in this role as the *practical* I, Fichte now argues that from it as the sole source flows the order of what *ought* to be, the order of the *ideal*. Confinement of this unending striving by an opposing principle, the not-I, leads to the order of the real; here the I becomes cognitive *intelligence*. Yet he says of this opposing force, of the not-I, that finite beings can feel but never know it. "All possible realizations which can occur of this force of the not-I for all times to come in our consciousness, the Philosophy of Science guarantees to derive from the defining powers of the I."

A geometric analogy will, I think, be helpful in clarifying the problem with which Fichte and Husserl are struggling, namely, to bridge the gap between immanent consciousness which, according to Heidegger's terminology, is ever-mine, and the concrete man that I am, who was born of a mother and who will die. The objects, the subjects, and the way an object appears to a subject, I model by the points, the coordinate systems, and the coordinates of a point with reference to a coordinate system in geometry. Relative to a system S of coordinates in a plane, consisting of three noncollinear points, there will be defined for each point p a triplet of number x_1, x_2, and x_3, with the sum equal to 1 (gravicentric coordinates). Here objects (points) and subjects (coordinate systems = triplets of points) belong to the same sphere of reality. The appearances of an object, however, lie in another sphere, in the realm of numbers.

Naive realism (or dogmatism, as Fichte calls this philosophical viewpoint) accepts the points as something which exists as such. Yet it is possible to build up geometry as an algebraic structure which makes use only of these number-appearances (modeling the experiences of pure consciousness). A point, so one defines it forthrightly, is simply a triplet of numbers x which add to 1; a coordinate system consists of three such triplets; algebraically, one explains how such a point p and such a system of coordinates S, determine three numbers ξ as the coordinates of p with reference to S. This triple ξ coincides with the triplet X which *defines* point p, if the system of coordinates S is the *absolute* one, which consists of the three triplets (1,0,0,), (0,1,0,), and (0,0,1,).

This coordinate system, therefore, corresponds to the absolute I, for which object and appearance coincide. In this argument, we never leave the sphere of numbers—or in the analogy, the immanent consciousness. After the fact, we can also do justice to the equivalence of all I's which must be required in the name of objectivity, by declaring that only such numerical relations are of interest as remain unchanged under passage from the absolute to an arbitrary coordinate system or, what amounts to the same, which remain invariant under an arbitrary linear transformation of the three coordinates. This analogy makes it understandable why the unique sense-giving I, when viewed objectively, i.e., from the standpoint of invariance, can appear as just one such subject among many of its kind. (Incidentally, a number of Husserl's theses become demonstrably false when translated into the context of this analogy—something which, it appears to me, gives serious cause for suspecting them.)

Beyond this, it is expected of me that I recognize the other I—the *you*—not only by observing in my thought the abstract norm of invariance or objectivity, but absolutely: you are *for you*, once again, what *I* am for *myself*: not just an existing but a conscious carrier of the world of appearances. This is a step which we can take in our geometric analogy only if we pass from the numerical model of point geometry to an axiomatic description. Now, we are not treating the points either as actual realities or as facts, nor have we established from the start an absolute coordinate system by identifying them with number triplets. Instead, the concept of point and the basic geometric relation, according to which a point p and a coordinate system S (i.e., triplet of points) determine a

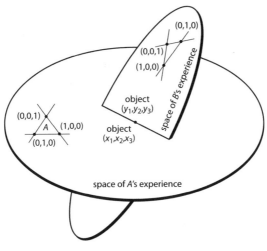

Figure 9.2

number triplet ξ, are introduced as undefined terms for which certain axioms are valid. This reveals that, above the viewpoints of naive realism and of idealism, it is possible to define, as a third one, a standpoint of transcendentalism which postulates a transcendental reality but which is satisfied with modeling it in symbols. This is the viewpoint to which the axiomatic construction of geometry corresponds in our analogy.

I do not wish to say that with this concept the enigma of selfhood is solved. Leibniz tried to resolve the conflict of human freedom and divine predestination, in that he allowed God to choose among the innumerable possibilities (for sufficient reasons) certain ones for existence, such as the beings Judas and Saint Peter, whose substantial nature thereafter determines their entire fate. The solution may be objectively adequate, but it is shattered by the desperate cry of Judas: Why did *I* have to be Judas! The impossibility of an objective formulation of this question strikes home, and no answer in the form of all objective insight can be given. Knowledge cannot bring the light that is I into coincidence with the murky, erring human being that is cast out into an individual fate.

At this point, perhaps, it becomes plain that the entire problem has been formulated up to now, and especially by Husserl, in a theoretically too one-sided fashion. In order to discover itself as intelligence, the I must pass, according to Descartes, through radical *doubt* and, according to Kierkegaard,

217

through radical *despair* in order to discover itself as existence. Passing through doubt, we push forward to knowledge about the real world, transcendentally given to immanent consciousness. In the opposite direction, however—not in that of the created works but rather in that of the origin—lies the transcendence of God, flowing from whence the light of consciousness—its very origin a mystery to itself—comprehends itself in self-illumination, split and spanned between subject and object, between meaning and being.

In a later stage of his philosophical development, Fichte went from idealism to a theological transcendentalism, as elaborated, for instance, in his *Guide to a Blessed Life*. The place of the absolute I is taken by God: "Being is thoroughly simple, not manifold, equal to itself, immutable and unchangeable; there is in it no genesis and no dissolution, no play or progression of forms, but forever the same peaceful Being and Existence." To be, as the revelation and expression of self-contained existence, is necessarily *consciousness* or *the conception of being*. Further, he says: "God not only exists, internal and hidden within himself, but He is *there*, and finds expression; yet his presence in its immediacy is necessarily *knowledge*, a necessity which can be seen in knowledge itself. . . . He is there, just as he is plainly within himself, without being changed in the transition from existence to presence, without an intervening cleft or separation. . . . And since that knowledge, or we ourselves are this divine presence, there cannot be any separation, differentiation, or cleavage in us either. Thus it must be, and it cannot be otherwise; therefore it is so." But then Fichte had to resort to sophistries and risky constructions, in order to be able to move from this unity of divine existence, which also pertains to the divine existence in us, to the manifold content of consciousness and the world.

I have spoken so far of the philosophers and the philosophical thoughts which stirred me in the period 1913 to 1922. Following Fichte, I pursued metaphysical speculations about God, the I, and the world for months on end. They seemed to unlock the final truth for me. I must confess, however, that every trace of them has flown from my memory.

All along, of course, there ran the stream which held a more central position in my life—mathematical research. This I shall pass over here, although a distorted picture will thus be given of the role that epistemological efforts and reflection played in my life. Only one thing should be mentioned: that in 1918 I established the first unified field theory of gravitation and electromagnetism. Although its basic principle, the "gauge invariance," has today been absorbed in a different form in quantum mechanics, the theory itself has long been surpassed by the modern development of physics which, alongside the electromagnetic field, has established the wave field of electrons and other elementary particles. At the same time that I was working on my 1918 theory, I was preoccupied with the foundations of mathematics, which are so closely related to the problem of infinity.

From the late works of Fichte I came upon Meister Eckhart, the deepest of the Occidental mystics. Despite his bonds with Plotinus and the conceptual apparatus of the Christian-Thomist philosophy, which he has at his command, one cannot doubt the originality of his basic religious experience: It is the inflow of divinity into the roots of the soul which he describes with the image of the birth of the "Son" or of the "Word" through God-Father. In turning its back on the manifold of existence, the soul must not only find its way back to this arch-image, but must break through it to the godhead that lives in impenetrable silence.

How freely Eckhart uses the words of scripture may be illustrated by the beginning of a Christmas sermon which has Matthew 2:2 as its text: "Where is he who is now born, the king of the Jews?" "Take notice first," he says, "where this birth took place: Yet, I claim, as often before, that this birth takes place in the soul in the same fashion as in eternity, not at all differently; for it is one and the same birth. And it truly takes place in the essence and the depths of the soul." The sermon closes with the words: "With this birth may God help us, who today has been born a new man, so that we poor children on earth will be born in Him as God; in this may He help us forevermore: Amen." Here speaks (the tone gives it away) a man of high responsibility and incomparably higher nobility than Fichte.

Of all spiritual experiences, the ones which brought me greatest happiness were the study of Hilbert's magnificent *Report on the Theory of Algebraic Numbers* as a young student in 1905, and in 1922 the reading of Eckhart, which captured me during a marvelous winter in the Engadine. Here I finally found for myself the entrance into the religious world, the lack of which had wrecked the beginning of a significant human relation ten years earlier.

However, my metaphysical-religious speculations, incited by Fichte and Eckhart, never achieved full clarity; this may perhaps also be due to the nature of the matter. In the following years, I undertook (among other things) to rethink the methodology of science, based on my scientific and philosophical experiences. Here my preoccupation with Leibniz became of increasing importance. After the soaring flight of the metaphysicist, there followed a period of soberness. What I had learned from the philosophers and thought through for myself found its deposit in the *Philosophy of Mathematics and Natural Sciences*, which was published in 1926. The drafting was done in a few weeks of vacation; but for the preceding year, I had been addicted to browsing in the literature of philosophy, making excerpts as I read—like a butterfly who, as it flies from flower to flower, endeavors to get a bit of honey from each one. An epistemological conscience, sharpened by work in the exact sciences, does not make it easy for the likes of us to find the courage for philosophical utterance. One cannot get by quite without compromise. Let me remain silent about that. The product of the struggle is available in print to anyone for whom this is of interest. All I wanted to describe here was the philosophical soil in which it is rooted.

At the same time I also attained a certain high point in mathematical research, exploring the structure of semisimple continuous groups. With this my development essentially came to a close. I don't know whether this is true for other people, too; but as I look back over my life, I find that the time from youth to an age of about thirty-five or forty, during which one's development steadily pushes forward into areas new to experience and understanding, appears incomparably richer than the following period of maturity and age. Of course in later years I did not remain unaffected either by the great revolution which quantum physics brought about in natural sciences, or by existentialist philosophy, which grew up in the horrible disintegration of our era. The first of these cast a new light on the relation of the perceiving subject to the object; at the center of the latter, we find neither a pure I nor God, but man in his historical existence, committing himself in terms of his own existence.

In 1930 I returned from Zurich to Göttingen as the successor to David Hilbert. When National Socialism broke over Germany in 1933, I emigrated to America, deeply revolted by the shame which this regime had brought to the German name. In the United States when I faced the task of getting out an English version of my old philosophy book, I no longer had the courage to write it anew in the light of the philosophical and scientific developments which had taken place in the meantime. I limited myself to applying an improving hand to the text, reworking several sections, and adding a number of appendices, whose preparation caused me more trouble than the original book had. All too often I considered giving up the effort completely or, when the manuscript of the appendices was finished, throwing it into the fire. The reason for these troubles and deliberations is perhaps best explained in a few lines from T. S. Eliot's "Four Quartets," which I used as the motto of the foreword:

Home is where one starts from. As one grows older
The world becomes stranger, the pattern more complicated
Of Dead and Living.

I am all the more thankful that, for this English edition, along with a later small book on symmetry which so far has been published only in English, I received the Arnold-Reymond Prize. To hold the lectures on "Symmetry" at Princeton had given me happiness.[10] At the time, I felt like a man who labored through a long day's work, doing his share as well as he could in the conflict of ideas and of human demands, and who now, as the sun is sinking and the conciliating night approaches, plays himself a quiet evening song on his flute.

And this concludes the rendering of my account.

Notes ▣

[All material in square brackets is by the editor; other notes are by Hermann Weyl. See the References for the abbreviations used below.]

Introduction

1. [*ECP* 8 (included as a supplement in vol. 10): 153–154 [93–94]. Einstein was writing in 1918 to Vero Besso, the son of his close friend Michele.]
2. [Atiyah 2002, 3. Penrose 1986, 23, calls Weyl "the greatest mathematician of this [twentieth] century."]
3. [Atiyah 2002, 13; see also Woit 2006, 118, 262–264.]
4. [In a 1919 letter to the physicist Arnold Sommerfeld, Weyl already emphasized that "I have always merely crossed into your field and never penetrated into your subject matter"; in 1949, Weyl wrote to Carl Seelig that "it is not correct that I 'changed to physics' upon my return to the ETH from German military service in 1915. I always remained a mathematician, also when I dealt with relativity or quantum theory." Translated in Sigurdsson 1991, 250–251.]
5. ["Insight and Reflection," 219, above. Weyl's comparison to the Pied Piper is in Weyl 1944, 132, his most extended memoir on his teacher. For Weyl's relation to Hilbert, see Reid 1996, 94–95, 104–105. For the Göttingen circle, see also Peckhaus 1990, Sigurdsson 1991, 1994.]
6. ["Princeton Bicentennial Address," 168, above. Sigurdsson 1991 gives a detailed description of Weyl's formative years and emphasizes his reaction to the Great War, here resonating with the thesis of Forman 1971 about the reaction of German physicists to their traumatic times. For a critical perspective, see Mancosu and Ryckman 2005.]

7. [Einstein's friend Marcel Grossmann was his principal mathematical adviser during this period leading to the 1916 formulation of general relativity; Einstein only began talking to Weyl about the implications of general relativity later. Weyl 1955b gives more details about his Zurich years.]

8. [This work went through many editions; the fourth German edition (1921) is available in translation as Weyl 1952a. For a detailed and helpful commentary, see the essays in Scholz 2001b. Another important early exposition was Pauli's 1921 book (1981), which considered Weyl's work on 192–202, as does Eddington 1923, 196–212. Einstein's quote about Weyl's book comes from *ECP* 8:669–670 [491]; for other comments of his about Weyl in this period, see 8:305 [225], 365–366 [265–266], 379–389 [277], 815 [598], 849 [622], 858–859 [629–630] ("I have confidence that Weyl not only is outstanding but also is a very delightful fellow personally as well. I'll not forego any opportunity to meet him."); 8 (as a supplement to 10): 160–163 [98–99]; from 1920, see 10:346–349 [215–218], 540–541 [340–341]. For the way Weyl's book "captivated" the young Enrico Fermi, see Holton 2005, 50–51.]

9. [*ECP* 8:710 [522]. Emphasis original.]

10. [For Weyl's recollection of his conversation with Willy Scherrer, see his "Princeton Bicentennial Address," 168, above. For the personal details of Weyl's circle, see Moore 1989, 149, 175–176, 275, 323; Anny's love affair with Weyl seems to have begun in their early Zurich days, by 1922 (149); her letters confirm its long duration (see her 1936 letter on 323).]

11. [Here Weyl was following the ideas and terminology of parallel transport given by Tullio Levi-Civita 1917, who assumed that Riemann's space is embedded in a Euclidean space of higher dimension. On Levi-Civita, see Reich 1992 and Bottazzini 1999; on the history of affine connections, see Scholz 1999a.]

12. [In this spirit, Galison 2003 investigates the connection between the growth of time synchronization (especially in relation to railways) and the development of Einstein's and Poincaré's conceptions of relativity. Yet both these men were much more involved in practical life (Einstein at the patent office) than Weyl, whose own writings do not use this railroad example, much more often referring to philosophical precedents. Weyl's original term for "gauge invariance" was *Maßstab-invarianz*, "measuring-rod invariance"; only later did he begin to use the term *Eichung*, which also means "gauge" in the railroad sense.]

13. [Weyl 1918d presents his further mathematical development of this "purely infinitesimal geometry"; see Scheibe 1988, Scholz 1995, 2000, 2001a, 2004a, and Sieroka 2007 for detailed discussion of Weyl's work and its relation to Fichte's philosophy. Weyl's work on this question also led to his treatment of the "problem of space" (*Raumproblem*), that is, to investigate the conditions necessary for the line element to have a "Pythagorean" metric such as Riemann originally assumed (namely, the quadratic form $ds^2 = \sum_{\mu\nu} g_{\mu\nu} dx^\mu dx^\nu$). This problem already had a long history in the work of Helmholtz and Lie (see Adler, Bazin, and Schiffer 1975, 9–16, Friedman 1995, Pesic 2007, 6–8, 47–70). Weyl's desire to reach a deeper level of understanding of this problem led him to make an important edition of

Riemann's original 1854 lecture, with commentary (Riemann 1919), and then to further mathematical work of his own (Weyl 1923) leading toward his conjecture that these "Pythagorean" metrics were singled out from all "Finsler" metrics (i.e., those with the more general form $ds = F(x_\mu, dx_\mu)$ that are homogeneous in the first degree in the dx_μ, meaning that $F(x_\mu, \lambda dx_\mu) = \lambda F(x_\mu, dx_\mu), \lambda > 0$) by the condition that for every such metric there is a uniquely determined affine connection (which specifies the parallel displacement of vectors). Though Weyl made considerable progress (Weyl 1923), his conjecture was proved by Detlef Laugwitz only in 1958; see Scholz 2004a for an excellent overview.]

14. [For a translation of Weyl 1918c, his first paper on his new theory, see O'Raifeartaigh 1997, 24–34, which includes Einstein's postscript and Weyl's replies, 34–37. For their further correspondence on this point, see *ECP* 8:709–711 [521–523], 720–729 [529–534], 741–745 [544–546], 824–825 [604–605], 877–881 [642–644], 893–895 [654–655], 948–949 [696], 954–957 [699–701], 966–969 [708–709], 971–972 [711]. For helpful accounts of Weyl's theory, see Adler, Bazin, and Schiffer 1975, 491–507, Pais 1982, 347–341, Vizgin 1994, 71–112, and Ryckman 2003, 2007, which connects Weyl's use of phenomenology to his 1918 theory. Ryckman 2003 gives a particularly full discussion of the status of Einstein's objection and Weyl's counterarguments, as does Fogel 2008, 45–120. For Weyl's concepts of "persistence" (*Beharrung*) and "adjustment" (*Einstellung*), see his use of these terms in "Electricity and Gravitation," 23–24, above. The use of light signals, rather than "rigid" measuring rods and clocks, was also advocated later by Ehlers, Pirani, and Schild 1972.]

15. [For the papers of Kaluza and Klein, see O'Raifeartaigh 1997, 44–76. Note that their fifth dimension is spatial, rather than temporal, in character; in comparison, Minkowski's fourth dimension was time.]

16. [For his speculation on matter as curved, empty space, see Weyl 1924b, which Wheeler 1988, 478–482, cites as proposing the essential thrust of his geometrodynamics (1962); the whole essay is a warm and interesting *hommage*. Adler, Bazin, and Schiffer 1975, 507–532, treats the Rainich-Misner-Wheeler theory further. For Faraday and Maxwell's views, see Pesic 1988–1989, 2002, 70–84. For Weyl's concepts of matter, see Scholz 2007.]

17. [For the Schrödinger (1922) and London (1927) papers, see O'Raifeartaigh 1997, 77–106; London's 1926 letter to Schrödinger is quoted in Moore 1989, 146–148, which gives a good brief account of Weyl's theory. Scholz 2008 describes Weyl's involvement with the "new" quantum theory. Kaluza and Klein already had noted that Weyl's concept of gauge, interpreted as a complex number, leads directly to a complex-valued (wave) function. In 1960, C. N. Yang pointed out that applying Weyl's theory to an electron would imply that an electron brought in a closed path back to its starting point would have an altered phase, which in fact is true, as pointed out in 1959 independently by Bohm and Aharonov and confirmed experimentally the following year; see O'Raifeartaigh 1997, 84–85.]

18. [For Weyl's relation to Schrödinger in this period, see Moore 1989, 191–220. Weyl's description of the "late erotic outburst" is cited by Pais 1986, 252; his gift inscription is quoted in Moore 1989, 291–292.]

19. [Weyl's *agens* theory of matter is excellently treated in Sieroka 2007, an extremely helpful paper that shows this theory's connection with Weyl's study of Fichte. See also Scholz 2007 for the development of Weyl's concept of matter. The 9 December 1919 letter to Pauli 1979, 1:6, is cited as translated by Sigurdsson 1991, 199. Weyl 1920b also presents his views of the fundamentality of the statistical aspect of physics, well discussed by Sigurdsson 1991, 180–185. It should be noted that many others around this time already shared the view that the discontinuous character of quantum phenomena implied that no deterministic classical theory would be sufficient; see, for instance, Pauli 1981, 205–206. For Schrödinger's and Weyl's shared interest in Eastern religions, see Moore 1989, 155–156.]

20. [Weyl 1924b, 510, gives his analogy between ego and matter, quoted here as translated in Sieroka 2007, 94. Weyl's 1920 letter to Felix Klein is quoted as translated by Sigurdsson 1991, 204. The quotation from Weyl 1923, 44, is also given in the translation by Skúli Sigurdsson in Scholz 2004a, 176. Wheeler 1988 clarifies the extent to which he knew these papers of Weyl's from the 1920s, making it even more plausible that he and Feynman (then his student) would have been influenced by them in Wheeler and Feynman 1949. Their theory was an important step in the development of quantum electrodynamics; see Schweber 1994, 380–389. Weyl 1924b, 510, brings up the issue of topology in fundamental physics. Sigurdsson 2007 emphasizes the importance of this "pre-Anglophone" phase of Weyl's work, and of the decades before the Second World War for the history of science in general, "a world of philosophical and literary discourse nourished by the German language."]

21. [Weyl's 1955 commentary comes from a postscript he wrote then for his 1918 paper; see O'Raifeartaigh 1997, 36–37. For Noether's theorem, see Tavel 1971 and Kosmann-Schwarzbach 2004; for the life of its creator, see Dick 1981, which contains Weyl's obituary of Noether on 112–152 (Weyl 1935); for its later influence, see Rowe 1999 and Brading 2002.]

22. [For a translation of Weyl's 1929 paper, see O'Raifeartaigh 1997, 107–144, which includes an excellent commentary. The Yang-Mills paper is available there on 182–196, as well as other important early papers in the development of gauge fields by R. Shaw and R. Utiyama. See also Yang 1986, Mielke and Hehl 1988, Pais 1986, 344–346, and Cao 1997.]

23. [This 1952 letter from Weyl to Carl Seelig is translated by Sigurdsson 1991, 253. See also Ryckman 2003 for discussion of this exchange of roles. Fogel 2008, 133–139, discusses Einstein's concept of theoretical "rigidity," which may render Einstein's later and earlier positions more consistent. See also Fogel's discussion of Weyl's methodology of science, 139–159.]

24. [Dieudonné 1976, 281. For Husserl's letters to Weyl (1918–1931), including Husserl's description of himself as "Exmathematicus" and also his desire to collaborate with a "philosophically gifted mathematician," see Van Dalen 1984 and Tonietti 1988. For the influence of philosophy on Weyl's work, see Peckhaus 1990; Sigurdsson 1991, 177–231; Mancosu 1998, 65–142; Coleman and Korté 2001; van Atten, van Dalen, and Tieszen 2002; Brading, 2002; Mancosu and Ryckman 2002, 2005; Martin 2003; Ryckman 2003; Ryckman 2005, 77–176; Redhead

2003; Scholz 2004a, 2005a, 2006; Ryckman 2007; Sieroka 2007. Beside her studies in philosophy, Helene Weyl also became the translator of José Ortega y Gasset
from Spanish into both English and German; her intellect and character, no less
than her striking physical beauty, make her a figure of singular interest in the circles
she and Hermann Weyl frequented.]

25. [For Aristotle's arguments against the actual infinite, see his *Physics* book 3, chapter 5.
Weyl's classic book *The Continuum* (1917) gives his most extended presentation of
his approach to this problem. For excellent discussions of this work, see Feferman
2000 and Scholz 2006. Hilbert was disturbed and even angered by this turn in
Weyl's thinking; see Reid 1996, 148–157. For Weyl's image of the Pied Piper, see
n. 5 above, commented on by Sigurdsson 1994, 357. For a general account of the
controversy about the foundations of mathematics, see Kuyk 1977; for Weyl's intuitionism, see van Dalen 1995, Coleman and Korté 2001, 315–372; for Brouwer,
see van Dalen 1999, 2005, and Mancosu 1998, which gives a helpful collection of
Weyl's and Brouwer's writings, with commentary.]

26. [Weyl 1949a (2009), 54. Weyl 1947 notes that "it is significant that Hilbert bases his
formalized mathematics on the practical manipulation of concrete symbols, tokens
drawn with chalk on blackboard or with pencil on paper, rather than on some 'pure
consciousness' and its data. The precise and exact symbolism of science, which in
itself is devoid of meaning, needs as its basis the common language, inexact but
pregnant with meaning. No understanding is possible for one unwilling to run the
risk of misunderstanding." See Feferman 1998, 249–283, Scholz 2000, and Feferman 2000, 187, which concludes that Weyl ended by being "a predicavist in the
sense that he was only going to deal with things that were introduced by definition,
but not an absolute predicavist in the sense that everything had to reduce to purely
logical principles—rather, a predicavist given the natural numbers."]

27. [The quotes about the impossibility of a pointlike Now come from his 1927 essay,
33, above. Sieroka 2007, 2008, points out that Alfred North Whitehead spoke
on the subject of time at the same session of the Harvard conference as Weyl (see
Brightman 1927) and wonders whether there might have been some interaction
between them, given Whitehead's own attempts to reconsider the concept of time,
though there is no evidence of such contact in the written record.]

28. [Weyl 1934a, 60, above; he also notes that "whereas the backward half of the world,
cut off by $t =$ constant, determines the whole, the interior of the backward light-
cone does not. That is to say, only after a deed is done can I know all its causal
premises." Faulkner's character Gavin Stevens is speaking in *Requiem for a Nun*
(1951); Faulkner 1994, 535.]

29. [The 9 December 1919 letter to Pauli 1979, 1:6, is cited as translated by Sigurdsson 1991, 199, the quotation from Weyl 1920b, 121–122, as translated in Sieroka
2007, 92–93.]

30. [For his summary presentation, see Weyl 1926, from which the quotes here are
drawn; Sigurdsson 2001, 18–29, discusses the context of this article. For an excellent review of Weyl's work on cosmology, see Goenner 2001. Adler, Bazin, and
Schiffer 1975, 309–317, gives details of Weyl's class of solutions to the linearized

field equations of general relativity; see also Penrose 1986, 38–46. Weyl 1922 (the fourth edition translated in Weyl 1952a, 274), mentions the possibility of closed timelike world lines, anticipating Gödel's 1949 solution containing such closed world lines. Weyl's 1923 calculation of a de Sitter radius for the universe, based on early redshift data, is in Appendix III to his 1923 edition of *Raum-Zeit-Materie*; see Scholz 2005b. For a review of the 1916–1918 debate between Einstein, de Sitter, Weyl, and Klein about the earliest relativistic cosmological models (and the problems of their being static or not, empty or not), see Goenner 2001 and *ECP* 8:351–357; for the development of Weyl's own point of view, see Bergia and Mazzoni 1999. For the "large-number hypothesis" and the context of recent cosmology, see also Weyl 1934b and Harrison 2000, 474–490 (who ascribes this hypothesis to Dirac 1937); see also below, n. 6 to *The Open World* for further details. Schwinger 1968, 1271–1272, gives his take on the question Weyl had already raised, "Does the quantum stabilize the cosmos?"]

31. [Schwinger 1988, 107. Pais 1986, 267, recounts that Giulio Racah "spent a full year studying Weyl's book during the isolation following his move from Florence to Jerusalem. That was all he needed to get started on his subsequent well-known work on complex atoms." For Slater's reaction to the "group pest," contrasted with Edward Teller's positive response, see Sigurdsson 1991, 235–238. For further discussion of Weyl's book in the history of the group-theoretic approach, see Mackey 1988, Speiser 1988, Coleman and Korté 2001, 271–314, and Scholz 2005a.]

32. [See Pais 1986, 346–352, Kragh 1990, 20–21, 42–43, 64–65, 90–91, and Schweber 1994, 66, for Dirac's initial proposal; Weyl 1950, 262–263, gives his argument that Dirac's holes cannot be protons but must have the same mass as electrons. Dirac 1971, 56–59, and 1977, 145, give his commentary. Weyl 1950 (written in 1930) anticipates *TCP*, whose history and later development is discussed by Yang 1986 and Coleman and Korté 2001, 293–311. For the relation between the *TCP* theorem and identicality, see Pesic 1993. Weyl 1952b (1980), 20–27, brings forward several of these symmetries in the context of art; though writing before the discovery of parity violation (1957), he does meditate on Heinrich Wölfflin's observation "that right in painting has another *Stimmungswert* than left" (23). Kragh 1990, 102–103, and Ryckman 2003, 84–84, discuss Dirac's turn toward mathematical formalism in response to Weyl's argument, a reaction also shared by Eddington 1923, 222–223, 237–240: "But we must not suppose that a law obeyed by the physical quantity necessarily has its seat in the world-condition which that quantity 'stands for'; its origin may be disclosed by unraveling the series of operations of which the physical quantity is the result. Results of measurement are the subject-matter of physics; and the moral of the theory of relativity is that we can only comprehend what the physical quantities *stand for* if we first comprehend what they *are*."]

33. [Michael Weyl used this phrase in a conversation about his father (private communication); his recollections in Chandrasekharan 1986, 95–100, 104–108, are both rich in remembered detail and delightful. The Yale lectures published in *The Open World* (1932) were originally delivered in 1930–1931.]

34. [For the situation of Hilbert and his circle in Göttingen, see Reid 1996, 203–215; regarding Weyl's return to Göttingen and his regrets, see Sigurdsson 1991, 263–277, who translates his 1933 letter acknowledging the danger to his family on 271. Sigurdsson 1996 reflects on the whole context of Weyl's emigration, quoting Courant's 1933 letter to Oswald Veblen about Weyl on 53. Weyl 1953b gives his later (and generally positive) account of the German universities as he knew them. By curious coincidence, another quite unrelated Hermann Weyl, a physician who emigrated to Buenos Aires, was the author of a story (*Der Epileptiker*, Ems: Verlag Kirchberger Presse, 1927) and edited and contributed to an anthology (*Maimonides*, Buenos Aires: Editorial Omega, 1956), is sometimes erroneously identified with the mathematician Hermann Weyl in bibliographies.]

35. [Quoted from Weyl 1934a, 96, above.]

36. [Quoted from Weyl 1934a, 113, above.]

37. [See Weyl 1947 and also n. 6 to "Man and the Foundations of Nature," below.]

38. [For Einstein's stubbornness, see Holton 1996, 180.]

39. ["Man and the Foundations of Science," 179, above. To some extent, this shorter (and hitherto unpublished) work stands in here for his longer published German essay for an Eranos conference, Weyl 1949b; the two works overlap so much that I decided to include this shorter statement, especially because Weyl wrote it in English. For another statement of Weyl's views about symbols in mathematics and physics, see Weyl 1953a.]

40. [Weyl 1952b (1980), which grew out of Weyl 1938, his Joseph Henry Lectures in Washington, DC. Penrose 1986, 24–38, reflects on Weyl's work on symmetry.]

41. [Weyl 1952b (1980), 17, 132 (on relativity theory and symmetry), 26 (on contingency), 64–65 (Hans Castorp's dream). On the individuality of snowflakes, see Pesic 2002.]

42. [This rare piece of journalism from the *Wisconsin State Journal*, April 31 [*sic*], 1929 ("P. A. M. issue"), entitled "Roundy Interviews Professor Dirac," is quoted in Schweber 1994, 18–20. Given the date, I surmise that Dirac might have been thinking about Weyl's arguments against his identification of antielectrons as protons (see n. 32, above), showing also that Dirac struggled with these arguments, at least for some time, before acceding to Weyl's point.]

1.   Electricity and Gravitation (1921)

[Weyl 1921a originally appeared in English as translated by Robert W. Lawson (here with slight corrections) and was an expanded version of a German paper, Weyl 1920a, passages from which are also translated below in the notes where they are significantly different or illuminating.]

1. [For the work of Mie and Lorentz, see Sigurdsson 1991, 107–114, and Vizgin 1994, 6–12, 26–38.]

2. [Weyl had coined the German term *affin zusammenhängend*, which has become widely used in English as "affinely connected"; his term "guiding field" (*Führungs-feld*) to describe the metric did not come into general use but gives an interesting insight into Weyl's understanding of the nature and efficacy of the metric considered as a field. Weyl 1952b (1980), 93–98, gives a beautiful discussion of affine geometry in the context of artistic symmetry.]

3. [The "line element" gives the invariant length between two events ds, such as the two ends of a measuring rod or two ticks of a clock. Note that Weyl seems to treat the metrical field as more fundamental than the affine connection; in later developments of unified field theories by Einstein and others, the affine connection was considered more fundamental and the metrical connections derived from it.]

4. [Weyl 1920a notes that "the metric ('the state of the field-ether') determines uniquely the affine connection (the 'gravitational field')."]

5. [Note that here Weyl emphasizes the fundamental role of infinitesimal displacement, as discussed above in n. 13 to the Introduction.]

6. [Weyl 1920a notes that "*Every* natural law allows itself to be formulated as coordinate invariant as well as gauge invariant; the principles of coordinate and gauge invariance will guide the formulation of physical laws first through the assumption that the laws in arbitrary coordinates and arbitrary gauge have a *simple mathematical form*."]

7. [Weyl uses the term *Wirkungsgrösse*, translated here as "magnitude of action," to denote the action function from which Maxwell's equations can be derived by minimizing that action (more precisely, by making it stationary). Such considerations of stationary action were important in Einstein and Hilbert's original presentations of the formalism of general relativity, carried forward by Weyl in his unified field theory (see Weyl 1918c).]

8. [By *Streckenübertragung* Weyl refers to the problem of the change of direction and (in general) of lengths of measuring rods transported parallel to themselves. As discussed in the Introduction, Einstein had objected that, in Weyl's theory, if one transports such a measuring rod in a closed curve, when it returns back to its starting point, its length would have changed, which means that its length depends on the history that led to the rod arriving there. This Einstein found unacceptable because he felt it would violate the observed constancy of spectral lines of various elements seen in astronomical objects throughout the visible universe.]

9. [For further discussion of these terms, see Fogel 2008, 53–88.]

2.   Two Letters by Einstein and Weyl on a Metaphysical Question (1922)

[Translated from Bovet 1922 (omitting or summarizing Bovet's introduction and final commentary) by the editor, with the essential help of Philip Bartok and Norman Sieroka. These letters do not appear in *WGA* or (as yet) in *ECP* and have not previously been republished or translated.]

1. [Spinoza had argued that physical events parallel attendant psychological states without there being any causal link between them; much later, Fechner carried this idea over into his discussion of the putative laws of stimulus and response.]

2. [It is not specified in this letter (or in Bovet's commentary) which passages from Poincaré Einstein exactly had in mind here.]

3. [For this "*agens* theory of matter," see Sieroka 2007, Scholz 2007, and the Introduction.]

4. [In his final paragraph, Weyl cites for further reading a number of general articles by Schottky 1921, Nernst 1922, and himself (Weyl 1920b).]

3. Time Relations in the Cosmos, Proper Time, Lived Time, and Metaphysical Time (1927)

[Translated from Weyl 1927 by the editor, with essential help from Philip Bartok and Norman Sieroka. This paper does not appear in *WGA*, and it has not been republished or translated previously.]

1. [Weyl's imagery of "fibers" was carried over into the concept of a fiber bundle (as a generalization of tangents to higher-dimensional curved spaces), which remain important for later mathematical treatments. For the history of fibration, see Zisman 1999. Fiber bundles were introduced into physics explicitly by Wu and Yang 1975, who compiled a "dictionary" translating between this mathematical terminology and that already in use by physicists.]

2. [Weyl is, of course, describing the now well-known way of visualizing space-time presented by Hermann Minkowski in his famous 1908 lecture "Space and Time," in Lorentz et al. 1923, 75–91, here in the context of general relativity.]

3. [Here again Weyl alludes to Leibniz's concept of monad, which here seems to indicate the primacy of the point-eye, of a consciousness whose active "creeping" is bringing the world line to life.]

4. [The "world tubes" (or perhaps *Weltröhren* should be rendered "world reeds") replace the earlier world lines of Minkowski, which imply pointlike events that Weyl has just argued must be considered to have some finite extension.]

4. The Open World (1932)

[Weyl 1932, comprising his Terry Lectures at Yale University in 1930–1931.]

1. [The closed three-dimensional space Weyl here is describing is now generally called a "3-sphere"; for a fascinating discussion of Dante in this light, see Peterson 1979.]

2. [The Greek word *apeiron* literally means "without limit," in the sense (for instance) that Euclid held that any magnitude could always be divided into smaller magnitudes. However, Euclid and Aristotle interpret this to mean a *potential* infinitude

of such divisions, rather than an *actual* infinitude of already-completed divisions. These correspond to the ancient and modern conceptions of infinity, simply put; Weyl will often refer to Georg Cantor's radical ideas about the infinite as the exemplar of the modern concept of the actual infinite.]

3. [The word "geodesic" denotes, in surveying, the line of least distance between two points on a curved surface; thus, great circles are the geodesics on a sphere. Following Riemann and Einstein, geodesics in curved spaces will replace ordinary "straight lines" on a Euclidean plane in the sense that each of them represents the straightest possible line in each case.]

4. [Weyl here refers to Hesiod, *Theogony* II:116: "First of all, the Void [*Chaos*] came into being, next broad-bosomed Earth, the solid and eternal home of all. . . ."]

5. [See Jammer 1966, 339–340, for a helpful account of the Stoic and later views on the nature of space and ether.]

6. [By "ether" throughout this section Weyl does not mean a space-filling substance (such as had been thought by Maxwell necessary as a medium through which light waves would propagate, for instance) but rather Weyl uses "ether" here to speak about empty space itself in the sense of its inertial properties. In fact, sometimes Weyl uses "ether" as a synonym for "field," presumably thinking of Einstein's gravitational field essentially as the description of space itself, considered now to have a certain physical reality and hence describable as "ether" in this new sense (see, for instance, Weyl 1952a, 311). Weyl, like Einstein, was inclined to think that local inertial properties of matter (and hence the space they moved in) were due to the influence of very distant matter, a view known as Mach's Principle (see Sciama 1969). If so, Weyl seems to imply (though without having expressed the argument fully in any place I can find in his works) that this inertial property of space should be expressed in terms of the "large-number hypothesis" (mentioned in n. 30 to the Introduction, above), more fully discussed in Harrison 2000, 474–490. Weyl 1926 mentions the primary "large number," the ratio of electric and gravitational forces between a proton and an electron, $N_1 = e^2/Gm_nm_e = 0.2 \times 10^{40}$, where e is the charge of the electron, G the Newtonian constant of gravity, m_n and m_e the masses of the proton and electron, respectively. Dimensionless numbers of the order of $N_1 \sim 10^{40}$ recur throughout cosmology in many different contexts, for instance as the ratio between the Hubble length ($L_H = c/H = 9.25 \times 10^{27} \, h^{-1}$ cm, where h is a parameter of order 1 dependent on observational data, not to be confused with Planck's constant) and the classical electron radius ($a = e^2/m_ec^2 = 2.82 \times 10^{-13}$ cm). In the present passage, Weyl notes that the ratio between the Planck length ($a^* = \sqrt{G\hbar/c^3} = 1.61 \times 10^{-33}$ cm) and the classical radius of an electron a is roughly 10^{20}. In Weyl 1952a, 262, the number 10^{20} also emerges as the value of e/m for an electron.]

7. [Here and in the following passage Weyl cites from Hölderlin's poems "Die Muße" and "An den Aether."]

8. [Carl Friedrich Georg Spitteler (1845–1924), a Swiss writer, won the Nobel Prize in 1919 for this visionary epic poem.]

9. [Bacon's phrase comes from "The Great Instauration" (1607): "Those, however, who aspire not to guess and divine, but to discover and know, who propose not to devise mimic and fabulous worlds of their own, but to examine and dissect the nature of this very world itself, must go to facts themselves for everything." (Bacon 1968, 4:28; Latin text 1:140).]

10. [Poincaré held that (for example) our choice whether to use Euclidean or non-Euclidean geometry to describe the physical world is purely a matter of convention; see Pesic 2007, 103–104, 144–146.]

11. [For the history of views about the sky's color, see Pesic 2005; though Weyl is correct that the density fluctuations are important consequences of atomic theory, the scattering takes place physically from the atoms, not the fluctuations (137–140).]

12. [Regarding the ergodic hypothesis, see Brush 1967, 168–182.]

13. [Max Born emphasized in 1926 that the wave function is completely determined by its wave equation, but when one calculates the absolute square of that wave function to find the probability of an event, that probability does *not* satisfy a partial differential equation and hence is only statistically observable; see Pais 1986, 258–267.]

14. [Instead of the Nicol plates Weyl refers to, polarizing filters, such as those used in polarized sunglasses, now are readily available; sheets of such Polaroid filters or, even more simply, a sheet of transparent plastic wrap, if stretched along a certain direction, will act as a (rather inefficient) polarizing filter.]

15. [Weyl here describes the Stern-Gerlach experiment, further illustrated and discussed in detail by Feynman et al. 1965, 3:1.1–1.10. For further discussion of the issue of identity in quantum theory with respect to Weyl's views, see Pesic 2002, 85–131, and French and Krause 2006, 127–131, 306–310.]

16. [For the work of Brouwer, see Benacerraf and Putnam 1964, 66–84, and Mancosu 1998, 1–63.]

17. [For the Mutakallimun or kālām, see Jammer 1960, 60–67, and Dhanani 1994.]

18. [Weyl has in mind the diagram in which one imagines B sliding along the curve toward A, so that the secant AB approaches the tangent at A, AD. To calculate the velocity at D, as Weyl explains following Newton, one takes the limit of the ratio

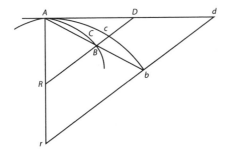

Figure N.1

AB:dt as *dt* → 0; see Newton 1999, 435–437 (*Principia*, book I, lemmas 6–8, from which the figure is taken).]

19. [For Galileo's construction, an early and prescient adumbration of the paradoxes of uncountability and of infinity, see Galileo 1974, 53–59 (referring to 92–97 of the standard page numbers).]

20. Following Leibniz, I place here the logical combination by "or" in analogy to the arithmetical + combination by "and" in analogy to ×.

21. [Much later the interpretation known as "quantum logic" considered quantum theory primarily a result of a new kind of logic, rather than a new physics.]

22. [Weyl gives a further commentary on Heidegger in "Man and the Foundations of Science," 189–193, above.]

5.   Mind and Nature (1934)

[Weyl 1934a comprises his Cooper Foundation Lectures at Swarthmore College in 1933.]

1. [Weyl 1934a includes at this point in the text the following bibliographical remark:] As regards literature I shall mention the following works: (1) As superb model examples of physical thinking—comprehensible also for the layman—Galileo's *Dialogo delli due massimi sistemi del mondo* and the *Discorsi delle nuove scienze*; in these writings modern natural science becomes conscious of itself for the first time and produces its first great achievements. (2) Among the philosophers none has penetrated as deeply into the nature of mathematical-physical thinking as Leibniz. I shall call attention to some of his works on occasion later on. (They are all to be found in the first volume of the *Hauptschriften zur Grundlegung der Philosophie*, edited in German by Duchenau and Cassirer in Vol. 107 of the *Meiner'sche Philosophische Bibliothek*.) (3) Of the more recent scientists with epistemological interest I shall mainly refer to Helmholtz. Outside of that I recommend the reading of E. Mach, particularly of his *Mechanics*. (4) From our generation there is the *Allgemeine Erkenntnislehre* (general theory of cognition) by M. Schlick and my contribution *Philosophie der Mathematik und Naturwissenschaft* (philosophy of mathematics and natural science) in the *Handbuch der Philosophie* [Weyl 1949a (2009)].

2. W. v. Scholz, "Der Spiegel," *Gedichte*; last lines of the prefixed motto.

3. [For further discussion of the divergences between hearing and seeing, see Pesic 2005, 161–162.]

4. [For the controversy about whether or not the eye is highly adapted to the solar spectrum, see Pesic 2005, 166–170, 241n14.]

5. [For the problem of violet (which Newton introduced somewhat arbitrarily as a fundamental spectral color) see Pesic 2005, 117, 161–166; for Newton's analogy between color and sound, see Pesic 2006.]

6. [In 1957, Edwin Land showed that full color photographs can be made from *two* black-and-white images taken through red and green filters, or through any two

colors separated by more than 30 nm in wavelength; see Land 1959. For a recent treatment of the visual system, see Shevelle 2003.]

7. [Accordingly, they and other insects whose visual spectrum spans an octave or more in wavelength can sense the phenomenon of recurrence (that middle C is somehow the "same note" as the c an octave higher).]

8. Helmholtz 1883, 2:656.

9. Helmholtz 1962, 433.

10. [Weyl refers to Lewis Carroll's poem in *Alice in Wonderland*: "You are old, Father William. . ."]

11. [Rudolf Hermann Lotze (1817–1881) advanced a theory of "local signs" to explain how positional information was transferred from the retina of the eye to the motor system. See Woodward 1978.]

12. [Though Weyl does not name him, here he describes Husserl's account of phenomenological reduction, as described (for instance) in Bell 1991, 161–171, 184–188, and by Ryckman 2003, 67–74.]

13. [See Schlick 1963.]

14. A famous line in Goethe's Faust: "In der Beschränkung zeigt sich erst der Meister." [Weyl's memory is in error; this line is actually from Goethe's poem "Das Sonnett," which begins "Natur und Kunst, sie scheinen sich zu fliehen."]

15. Addition of momenta is performed after the same "rule of parallelograms" as addition of forces: $I = I_1 + I_2$ in fig. 5.5.

16. Scholium, following the definition of the first book of the *Principia*.

17. [It should be noted that this so-called "ultraviolet catastrophe" was not known to Max Planck and first came to general attention after 1911; see Klein 1960.]

18. [See above, n. 11 to *The Open World*.]

19. [See above, n. 14 to *The Open World*.]

6.   Address at the Princeton Bicentennial Conference (1946)

[Transcribed from Weyl 1946, his typewritten manuscripts (in English) in the ETH Archive, Nachlass Hermann Weyl, which contains two versions of this address, the first (Hs 91a:17) marked (in Weyl's hand) "Dec. 1946. Bicentennial conference. First version [not used]"; the second (Hs 91a:18) bears the typed heading "(Bicentennial conference)." Despite Weyl's notations, though, the first version is longer and more ample, while the second is considerably shorter. In preparing the text printed here, I basically used the second version, with salient additions from the first version, even though Weyl omitted them from the second, if they helpfully amplify his meaning and are not too technical. A detailed list following the notes below specify these added passages, as well as readings from the first version not incorporated in the main text of this anthology (keyed to paragraph numbers I have inserted in the margins). For the political context of this conference, see Schweber 2000, 4–5, 10–11; for an abstract of the proceedings, see *Problems of Mathematics* 1946. For a masterly survey of Weyl's mathematical work, see Chevalley and Weil 1957 and (more briefly) Dieudonné 1976, Coleman and Korté 2001, 165–197, Atiyah 2002.]

1. [The standing waves of a vibrating string are examples of "eigenfunctions," the "characteristic functions" that are the possible states allowed by a differential (or partial differential) equation; the "eigenvalue" is (for instance) the frequency of an allowed standing wave. By the "gravest eigentone," Weyl means the fundamental frequency allowed by that membrane, the lowest pitch it sounds. See Yang 1986, 7–9.]

2. [Already from the end of the nineteenth century, physicists such as Kirchhoff had proved that the thermal distribution inside a heated oven (idealized as perfectly black so that it would absorb and radiate all frequencies equally) is independent of its material and size. In 1900, to explain the first accurate experimental measurements of such a blackbody, Max Planck hypothesized that the radiation in the blackbody could only take on discrete, quantized values according to his famous formula $E = h\nu$. Hence Weyl's work on this topic addressed an important physical system. See Sigurdsson 1991, 36–53, for a very helpful account of this phase of Weyl's work in the context of contemporary physics; Klein 1960 discusses the prehistory of the "ultraviolet catastrophe."]

3. [Weyl himself thought this his greatest achievement in mathematics, which was the basis of his famous application of group theory to quantum mechanics, detailed in Weyl 1950. The theory of groups characterizes and categorizes symmetries; for instance, the theory of permutation groups can characterize the symmetries of Platonic solids such as the cube or the icosahedron (see Pesic 2003). Sophus Lie had generalized these discrete symmetries to continuous symmetries, now known as Lie groups. For instance, the continuous rotational symmetries of a sphere are governed by the group of orthogonal (distance-preserving) rotations in three dimensions.]

4. [For Hodge's theory, see Weyl 1943.]

5. [In 1770, Edward Waring proposed that for every natural number n there is some integer s such that n is the sum of at most s kth powers of natural numbers. For instance, Lagrange showed in 1770 that every number could be expressed as the sum of no more than four squares; in 1909 Hilbert demonstrated the theorem for all k.]

6. [This result, often referred to now as "Weyl's sums," concerns sequences of real numbers $a_1, a_2, \ldots$, in the unit interval $[0, 1)$, which are "equidistributed mod 1" if, for every subinterval I in that unit interval, the number of those $a_1, a_2, \ldots, a_n$ for which the fractional part (a_i) lies in I tends, after being divided by n, to the length of I. See Narkiewicz 1988.]

7. [By "transcendency proofs," Hilbert means proofs that a certain quantity (such as π or e) is transcendental, meaning that it is not the solution of any finite-order algebraic equation with integer coefficients.]

8. [For Mie's theory, see Vizgin 1994, 26–38, and also Corry 2004.]

9. [Weyl alludes to Hilbert's famous remark that "no one will drive us from the paradise which Cantor created for us," rejecting those who (like Brouwer and Weyl also) were critical of the way Cantor introduced his transfinite number by means of actual infinite processes, which Weyl often criticizes in this volume.]

10. [Weyl's first book (1913) was a seminal exposition of the general concept of a Riemann surface, a deformed version of the complex plane whose topology can represent multivalued complex functions by showing their multiple values on separate "sheets" of the surface.]

11. [Lancelot Hogben (1895–1975) was a zoologist and medical statistician (also a pacifist and Marxist) whose book *Mathematics for the Millions* (1936) became a best-selling popularization of mathematics.]

12. [This is the end of the second version of this address; the remainder printed here is the conclusion of the first version.]

13. [Goethe, *Torquato Tasso*, V.ii.95–96; Weyl gives his own translation in the brackets.]

[Unless otherwise noted, the following list indicates sentences drawn from Weyl's first version (V1) that I have added to the second version (V2) to make the composite text printed in the main text of this volume, listed below by paragraph number of that text. I have also included below some passages from V1 not included in the composite text, indicated by an asterisk (*)]:

¶1: "The balance . . . of youth:"

¶2: [only in V1.]

¶¶3–4: "who combined . . . David Hilbert,"; "But slow travel . . . become concrete."; "Bernoulli's heuristic . . . highest honor" [replacing shorter version in V2.]

¶5: [V2 ends with:] "After the First World War, Courant improved my procedure. But it was by an entirely different and more powerful method that much later Carleman tackled the corresponding problem for eigenfunctions."

¶6: "Our work had just been completed . . . almost periodic functions": [*V1 ends instead with:] "When Peter and I had just completed our work, Harald Bohr visited Zurich and gave an inspiring talk on almost periodic functions. In the ensuing discussion I urged the desirability of proving the completeness relation for Fourier series without going through approximation by finite trigonometric sums—and failed to see at that moment that our method did just that when applied to the group of rotations of a circle. Thereby it also opened the way to a new treatment of almost periodic functions."

¶9: [*V1 adds at end of paragraph:] "Hilbert showed how to derive the limiting potential from the Dirichlet principle that he had put on a safe basis before. Koebe himself invented convergent processes of a more constructive character for the problems of conformal mapping, in particular for the mapping of schlicht domains, and *Verzerrungssätze* of a surprisingly simple and incisive character were derived by him and his collaborators. Many ramifications lead from here to the function-theoretic work of later years. A particularly fascinating problem taken up by the Finnish school is the type problem arising from the fact that there are two conformally anisomorphic open simply connected surfaces, the plane and the interior of the circle."

[*only in V1 then follows:] "The second outstanding event in function theory that left an indelible imprint on my mind was the appearance of Rolf Nevanlinna's paper on *meromorphic functions* in the *Acta Mathematica* 1925. The theory of entire functions had had its great success, proof of the asymptotic law for the distribution of prime numbers, before my time. Up to 1925 it had remained on the whole within the confines staked out by the early work summarized in Borel's well-known book of 1900. It had given inspiration for many charming, even profound, investigations. But Nevanlinna's new approach culminating in the defect relations, was like the turn of a mountain road that all of a sudden opens up to you a wide landscape. I must be content to record this impression; except on its outer edges— meromorphic curves—my own work has hardly ever touched these fields. In the hands of Rolf and Fritjof Nevanlinna, Ahlfors and many others, the new approach has proved extremely fruitful."

¶10: [*V1 includes before last sentence:] "The applications of topology have multiplied. One of the most important is the use of the fixed point theorems for the solution of non-linear differential equations."

¶11: [*V1 adds after second sentence:] "Hardy and Littlewood obtained asymptotic laws for the number of representations, the most interesting feature of which is the singular series." [*V1 adds at end:] "He even succeeded in solving corresponding problems where the variables range over the prime numbers instead of all integers. Another branch of the same tree was transplanted to this country by Rademacher."

¶12: "I shall . . . big splash." [*V1 includes then:] "[Frobenius' work on representations of groups does not fall within the period under review.]"

¶14: "a strange and superb malice of the Creator,"; "Good craftsmanship . . . not everything"

¶15: [only in V1.]

¶16: [as in V1, except for V2:] "The birth of quantum mechanics . . . mathematical event." [Before this, V2 has:] "But reviendrons à nous moutons [let us return to our sheep, i.e., let us return to our point]."

¶18: "In an example . . . greater accuracy."; "I recommend . . . boundless abstraction."

¶20: [only in V1.]

¶21: [where V1 reads:] "truth about the world that is and truth about our existence in the world" [V2 has:] "basic truth and ideas that bear the mark of inevitability."

¶23–end: [only in V1.]

7. Man and the Foundations of Science (ca. 1949)

[This previously unpublished manuscript in the ETH Archiv (Hs 91a:28) overlaps considerably with Weyl 1949b, his Eranos-Jahrbuch contribution, and accordingly I infer that this manuscript's date is also around 1949, though there are no indications in it or elsewhere in Weyl's *Nachlass*.]

1. [Weyl writes that there is "no exchange of momentum" between the colliding atoms in the plane tangential to their impact but clearly means no *net* exchange; the total momentum in that plane must remain zero if initially assumed to vanish.]
2. [Fichte, "Bestimmung des Menschen ...," 137, in Fichte 1971, 2:229.]
3. [*Res cogitans*, the "thinking thing," is Descartes' term for mind, opposed to matter as "extended thing."]
4. [For Lotze and his "local signs," see above, n. 11 to *Mind and Nature*. Weyl's manuscript has "graduations," which I take as a misprint for "gradations."]
5. [In this now-famous term, Heidegger appropriated the ordinary word for "presence," literally "being-there" (*Da-sein*), to indicate the "existential" crux of human being.]
6. [Weyl 1947, 1 (published in the records of the discussions in the symposium on *Problèmes de Philosophie des Sciences* 1947, fascicle 1, 30) emphasizes this point: "In the times of Democritus, Descartes and Huyghens one could still believe that physics revealed the true real world, atoms moving in space, behind the sensual and subjective appearances that, e.g., colored light beams were in reality oscillations of an ethereal fluid. All this is gone; intuitive space and time belong no less than the sensations on the subjective side of the ledger. On the other hand, it makes little sense to claim an array of symbols as the true objective world. Where then is the reality to which our constructions ultimately refer? The *Sinnesempfindungen*, as Mach thought? I think many philosophers are now willing to agree that sensations, far from being given in their naked purity, are theoretical abstractions, while the true raw material and the only firm ground on which to stand is the manifest commonplace world of our everyday life, in which we handle chair and table, go to a meeting, call on a friend, etc. There is no substitute for the understanding by which this world (in all its complexity) is disclosed to us."]

8. The Unity of Knowledge (1954)

[Weyl 1954, comprising his address at the Columbia University Bicentennial celebration.]

1. [See Weyl 1940.]
2. [For Weyl and Cassirer, see Röller 2002.]
3. [The German word *Wesensschau* is one of the key concepts of phenomenological philosophy, commonly translated as "intuition of pure essences."]
4. [Weyl's "alas! also philosophy" sounds like an ironic echo or rewriting of the famous opening of Goethe's *Faust*, Part I, lines 353–355: "Nun habe ich Religion, Jurisprudenz / Und leider auch Philosophie durchaus studiert. . ." (Now I have diligently studied religion, jurisprudence, and alas! also philosophy. . .).]

9.   Insight and Reflection (1955)

[Weyl 1955a, a lecture given at the University of Lausanne, Switzerland, in May 1954, here as translated by Howard Schindler in Saaty and Weyl 1969, 281–301. Note that figures 9.1 and 9.2 did not appear in the original German edition but were added in this English version, which was supervised by Joachim Weyl.]

1. [Weyl 1931, 18, which he expressed again at 81, above.]
2. [By "hylic" Weyl refers to the Greek term used by Aristotle, *hylē*, which denotes primal matter in the sense of the material out of which everything is formed, the "lumber" from which all natural beings are formed.]
3. [For Helmholtz's writings, see Helmholtz 1977, 39–71, Pesic 2007, 47–70.]
4. [For Riemann's famous inaugural lecture "On the Hypotheses that Lie at the Foundations of Geometry" (1854), see Pesic 2007, 23–40.]
5. [This was Weyl's well-known work on the so-called *Raumproblem*, to demonstrate the exact mathematical requirements that underlie the form of the Riemannian metric; see n. 13 to the Introduction and Weyl 1923 (1963).]
6. [Fritz Medicus was a professor of philosophy in Zurich, with whom Weyl studied and discussed Fichte; see Sigurdsson 1991, 221–223; Scholz 2001a, 2004, and Sieroka 2007 discuss the influence that Medicus's approach to Fichte had on Weyl.]
7. [Husserl used the term *epochē* to describe the suspension of judgment regarding the true nature of reality.]
8. [In Husserlian phenomenology, the "noetic" refers to the form of a conscious act (whether it is believed or doubted, for instance, rather than the content of what is believed or doubted, which Husserl called "noematic"). *Morphē* is the Greek philosophical term for form or shape.]
9. [Fichte 1984, 28–29.]
10. [Weyl 1952b (1980).]

References 🔳

[Note that the abbreviation *WGA* refers to Weyl's *Gesammelte Abhandlungen* (Weyl 1968), whose pagination is used for all references to those works. *ECP* refers to Einstein's *Collected Papers* (Einstein 1987–); in multivolume works, 2:111 refers to volume 2, page 111. The listing of editions below is not exhaustive; only the original (first) and most recent (English) editions are listed.]

Adler, Ronald, Maurice Bazin, and Menahem Schiffer. 1975. *Introduction to General Relativity*, 2nd ed. New York: McGraw-Hill.

Ash, Mitchell G., and Alfons Söllner, eds. 1996. *Forced Migration and Scientific Change: Emigré German-Speaking Scientists and Scholars after 1933*. Cambridge: Cambridge University Press.

Atiyah, Michael. 2002. "Hermann Weyl: November 9, 1885–December 9, 1955." *Biographical Memoirs of the National Academy of Sciences* 82:1–13.

Ayoub, Raymond B., ed. 2004. *Musings of the Masters: An Anthology of Mathematical Reflections*. Washington, DC: Mathematical Association of America.

Bacon, Francis. 1968. *The Works of Francis Bacon*, ed. J. Spedding. New York: Garrett Press.

Bell, David. 1991. *Husserl*. London: Routledge.

Bell, J. L. 2000. "Hermann Weyl on Intuition and the Continuum." *Philosophia Mathematica* 8:259–273.

Benacerraf, Paul, and Hilary Putnam, eds. 1964. *Philosophy of Mathematics: Selected Readings*. Englewood Cliffs, NJ: Prentice-Hall.

Bergia, Silvio, and Lucia Mazzoni. 1999. "Genesis and Evolution of Weyl's Reflections on de Sitter's Universe." In Renn et al. 1999, 325–343.

Borel, Armand. 2001. *Essays in the History of Lie Groups and Algebraic Groups*. Providence, RI: American Mathematical Society.

Bottazzini, Umberto. 1999. "Ricci and Levi-Civita: From Differential Invariants to General Relativity." In Gray 1999b, 241–259.

Bovet, E. 1922. "Die Physiker Einstein und Weyl antworten auf eine metaphysische Frage." *Wissen und Leben* 19:901–906.

Brading, Katherine. 2002. "Which Symmetry? Noether, Weyl, and Conservation of Electric Charge." *Studies in the Philosophy of Modern Physics* 33:3–22.

Brading, Katherine, and Harvey R. Brown. 2003. "Symmetries and Noether's Theorem." In Brading and Castellani 2003, 89–109.

Brading, Katherine, and Elena Castellani, eds. 2003. *Symmetries in Physics: Philosophical Reflections.* Cambridge: Cambridge University Press.

Brightman, Edgar Sheffield, ed. 1927. *Proceedings of the Sixth International Congress of Philosophy. Harvard University, Cambridge, Massachusetts, United States of America. September 13–17, 1926.* Toronto: Longmans, Green and Co.

Brush, Stephen G. 1967. "Foundations of Statistical Mechanics 1845–1915." *Archive for History of Exact Sciences* 4:145–182.

Cantor, G., and M. Hodge, eds. 1981. *Concepts of the Ether.* Cambridge: Cambridge University Press.

Cao, Tian Yu. 1997. *Conceptual Development of 20th Century Field Theories.* Cambridge: Cambridge University Press.

Chandrasekharan, K., ed. 1986. *Hermann Weyl: 1885–1985; Centenary Lectures delivered by C. N. Yang, R. Penrose, and A. Borel at the Eidgenössische Technische Hochschule Zürich.* Berlin: Springer.

Chatterji, S. D., ed. 1995. *Proceedings of the International Congress of Mathematicians, Zürich, Switzerland 1994.* Basel: Birkhäuser.

Chevalley, Claude, and André Weil. 1957. "Hermann Weyl (1885–1955)." *L'Enseignement* 3(3):157–187; *WGA* 4:655–685.

Chikara, Saski, Sugiura Mitsuo, and Joseph W. Dauben, eds. 1994. *The Intersection of History and Mathematics.* Basel: Birkhäuser.

Coleman, Robert, and Herbert Korté. 2001. "Hermann Weyl: Mathematician, Physicist, Philosopher." In Scholz 2001b, 161–388.

Corry, Leo. 1999. "Hilbert and Physics (1900–1915)," in Gray 1999b, 145–188.

———. 2004. *David Hilbert and the Axiomatization of Physics (1898–1918). From* Grundlagen der Geometrie *to* Grundlagen der Physik. Dordrecht: Kluwer.

Earman, J. S., C. N. Glymour, and J. J. Stachel, eds. 1977. *Foundations of Space-Time Theories.* Minneapolis: University of Minnesota Press.

Deppert, W., ed. 1988. *Exact Sciences and Their Philosophical Foundations.* Frankfurt am Main: Peter Lang.

Dhanani, Alnoor. 1994. *The Physical Theory of Kalām.* Leiden: E. J. Brill.

Dick, Auguste. 1981. *Emmy Noether 1882–1935*, tr. H. I. Blocher. Boston: Birkhäuser.

Dieudonné, Jean. 1976. "Weyl, Hermann." In *Dictionary of Scientific Biography*, ed. C. G. Gillispie, 14:281–285.

Dirac, P.A.M. 1937. "The Cosmological Constants." *Nature* 139:323.

———. 1971. *The Development of Quantum Mechanics.* New York: Gordon and Breach.

———. 1977. "Recollections of an Exciting Era." In Weiner 1977.

Eddington, A. S. 1923. *The Mathematical Theory of Relativity*. Cambridge: Cambridge University Press. (Reprint, 1963)

Ehlers, J., F.A.E. Pirani, and A. Schild. 1972. "The Geometry of Free Fall and Light Propagation." In O'Raifeartaigh 1972, 63–84.

Ehlers, Jürgen. 1988. "Hermann Weyl's Contribution to the General Theory of Relativity." In Deppert 1988, 83–105.

Einstein, Albert. 1987–. *The Collected Papers of Albert Einstein*. Princeton: Princeton University Press. [Note that the page references for the translations in the companion volumes for each volume of main text are included in square brackets.]

Ewald, William, ed. 1996. *From Kant to Hilbert: A Source Book in the Foundations of Mathematics*. Oxford: Clarendon Press.

Faulkner, William. 1994. *Novels 1942–1954*. New York: Library of America.

Feferman, Solomon. 1998. *In the Light of Logic*. New York: Oxford University Press.

———. 2000. "The Significance of Weyl's Das Kontinuum." In Hendricks et al. 2000, 179–194.

Feist, R. 2002. "Weyl's Appropriation of Husserl's and Poincaré's Thoughts." *Synthèse* 132:273–301.

Ferreiros, Jose, and Jeremy Gray, eds. 2006. *Architecture of Mordern Mathematics: Essays in History and Philosophy*. New York: Oxford University Press.

Feynman, Richard P., Robert Leighton, and Matthew Sands. 1965. *The Feynman Lectures in Physics*. Reading, MA: Addison-Wesley.

Fichte, Johann Gottlieb. 1971. *Fichtes Werke*, ed. Immanual Hermann Fichte. Berlin: W. de Gruyter.

———. 1984. *Versuch einer neuen Darstellung der Wissenschaftlehre (1797/98)*, ed. Peter Baumanns. Hamburg: Meiner Verlag.

Fogel, D. Brandon. 2008. "Epistemology of a Theory of Everything: Weyl, Einstein, and the Unification of Physics." Ph.D. diss., University of Notre Dame.

Forman, Paul. 1971. "Weimar Culture, Causality, and Quantum Theory 1918–1927: Adaptation by German Physicists and Mathematicians to a Hostile Intellectual Environment." *Historical Studies in the Physical Sciences* 3:1–116.

Frei, Gunther, and Urs Stammbach. 1992. *Hermann Weyl und die Mathematik an der ETH Zurich, 1923–1930*. Basel: Birkhäuser.

French, Steven, and Décio Krause. 2006. *Identity in Physics: A Historical, Philosophical, and Formal Analysis*. Oxford: Clarendon Press.

Friedman, Michael. 1973. "Grünbaum on the Conventionality of Geometry." In Suppes 1973, 217–233.

———. 1995. "Carnap and Weyl on the Foundations of Geometry and Relativity Theory." *Erkenntnis* 42:247–260.

———. 2001. *Foundations of Space-Time Theories: Relativistic Physics and Philosophy of Science*. Princeton: Princeton University Press.

Galilei, Galileo. 1974. *Two New Sciences*, tr. Stillman Drake. Madison: University of Wisconsin Press.

Galison, Peter. 1979. "Minkowski's Space-Time: From Visual Thinking to the Absolute World," *Historical Studies in the Physical Sciences* 10:85–121.

Galison, Peter. 2003. *Einstein's Clocks, Poincaré's Maps*. New York: W. W. Norton.

Gavroglu, Kostas, Jean Christianidis, and Efthymios Nicolaidis, eds. 1994. *Trends in the Historiography of Science*. Dordrecht: Kluwer Academic.

Gavroglu, Kostas, and Jürgen Renn, eds. 2007. *Positioning the History of Science*. Dordrecht: Springer.

Giere, Ronald N., and Alan W. Richardson, eds. 1996. *Origins of Logical Empiricism. Minnesota Studies in the Philosophy of Science*, vol. 16. Minneapolis: University of Minnesota Press.

Gödel, Kurt. 1949. "An Example of a New Type of Cosmological Solution of Einstein's Field Equations of Gravitation." *Reviews of Modern Physics* 21:447–450.

Goenner, Hubert. 2001. "Weyl's Contributions to Cosmology." In Scholz 2001c, 105–137.

Grattan-Guinness, Ivor, ed. 2005. *Landmark Writings in Western Mathematics, 1640–1940*. Amsterdam: Elsevier.

Gray, Jeremy. 1999a. "Geometry—Formalisms and Intuitions." In Gray 1999b, 58–83.

———. ed. 1999b. *The Symbolic Universe: Geometry and Physics 1890–1930*. Oxford: Oxford University Press.

Greenspan, Nancy Thorndike. 2005. *The End of the Certain World: The Life and Science of Max Born*. New York: Basic Books.

Grene, Marjorie, and Debra Nails, eds. 1986. *Spinoza and the Sciences*. Dordrecht: D. Reidel.

Grünbaum, Adolf. 1973. "On the Ontology of the Curvature of Empty Space in the Geometrodynamics of Clifford and Wheeler." In Suppes 1973, 268–295.

Harrison, Edward. 1987. *Darkness at Night: A Riddle of the Universe*. Cambridge, MA: Harvard University Press.

———. 2000. *Cosmology: The Science of the Universe*. 2nd ed. Cambridge: Cambridge University Press.

Hawkins, Thomas. 1999. "Weyl and the Topology of Continuous Groups." In James 1999, 169–198.

———. 2000. *Emergence of the Theory of Lie Groups. An Essay in the History of Mathematics 1869–1926*. Berlin: Springer.

Helmholtz, Hermann von. 1883. *Wissenschaftliche Abhandlungen*. Leipzig: Johann Ambrosius Barth.

———. 1954. *On the Sensations of Tone as a Physiological Basis for the Theory of Music*, 2nd ed, tr. Alexander J. Ellis. New York: Dover.

———. 1962. *Helmholtz's Treatise on Physiological Optics*, ed. James P. C. Southall. New York: Dover.

———. 1968. *Über Geometrie*. Darmstadt: Wissenschaftliche Buchgesellschaft.

———. 1971. *Selected Writings of Hermann von Helmholtz*, ed. Russell Kahl. Middletown, CT: Wesleyan University Press.

———. 1977. *Epistemological Writings*, tr. Malcolm F. Lowe, ed. Robert S. Cohen and Yehuda Elkana. Dordrecht: D. Reidel.

Hendricks, Vincent F., Stigandur Pedersen, and Klaus F. Jørgensen, eds. 2000. *Proof Theory: History and Philosophical Significance*. Dordrecht: Kluwer.

Hendricks, Vincent F., Klaus F. Jørgensen, Jesper Lützen, and Stig A. Pedersen, eds. 2007. *Interactions: Mathematics, Physics and Philosophy, 1860–1930*. Dordrecht: Springer.

Hesseling, Dennis. 2003. *Gnomes in the Fog. The Reception of Brouwer's Intuitionism in the 1920s*. Basel: Birkhäuser.

Holton, Gerald. 1996. *Einstein, History, and Other Passions*. Reading, MA: Addison-Wesley.

———. 2005. *Victory and Vexation in Science: Einstein, Bohr, Heisenberg, and Others*. Cambridge, MA: Harvard University Press.

James, I. M., ed. 1999. *History of Topology*. Amsterdam: Elsevier.

Jammer, Max. 1960. *Concepts of Space*. New York: Harper.

———. 1966. *The Conceptual Development of Quantum Mechanics*. New York: McGraw-Hill.

Joas, C., C. Lehner, and J. Renn, eds. 2008. "HQ–1: Conference on the History of Quantum Physics (Berlin July 2–6, 2007)." Berlin: MPI History of Science Preprint 350.

Klein, Martin J. 1960. "Max Planck and the Beginnings of Quantum Theory." *Archive for History of Exact Sciences* 1:459–479.

Kosmann-Schwarzbach, Yvette. 2004. *Les théorèmes de Noether: invariance et lois de conservations au XXe siècle*. Palaisseau: Éditions de l'École Polytechnique.

Kragh, Helge. 1981. "The Genesis of Dirac's Relativistic Theory of the Electron." *Archive for History of Exact Sciences* 24:31–67.

———. 1990. *Dirac: A Scientific Biography*. Cambridge: Cambridge University Press.

Kuyk, Willem. 1977. *Complementarity in Mathematics*. Dordrecht: D. Reidel.

Land, Edwin H. 1959. "Experiments in Color Vision." *Scientific American* 200(5): 84–99.

Laugwitz, Detlef. 1958. "Über eine Vermutung von Hermann Weyl zum Raumproblem." *Archiv der Mathematik* 9:128–133.

Leupold, R. 1960. "Die Grundlagenforschung bei Hermann Weyl." Ph.D. diss., Universität Mainz.

Levi-Civita, Tullio. 1917. "Nozione di parallelismo in una varietà qualunque e conseguente specificacione geometrica della curvatura Riemanniana." *Rendiconto del Circolo Mathematico di Palermo* 42:173–205.

Lorentz, H. A., A. Einstein, H. Minkowski, and H. Weyl. 1923. *The Principle of Relativity: A Collection of Original Memoirs on the Special and General Theory of Relativity*. New York: Dover.

MacAdam, David L., ed. 1970. *Sources of Color Science*. Cambridge, MA: MIT Press.

Mach, Ernst. 1906. *Space and Geometry in the Light of Physiological, Psychological and Physical Inquiry*, tr. Thomas J. McCormack. Chicago: Open Court.

Mackey, George. 1988. "Hermann Weyl and the Application of Group Theory to Quantum Mechanics." In Deppert 1988, 131–160.

Mancosu, Paolo. 1998. *From Brouwer to Hilbert. The Debate on the Foundations of Mathematics in the 1920s*. Oxford: Oxford University Press.

Mancosu, Paolo, and Thomas Ryckman. 2002. "Mathematics and Phenomenology: The Correspondence between O. Becker and H. Weyl." *Philosophia Mathematica* 10:130–202.

———. 2005. "Geometry, Physics and Phenomenology: Four Letters of O. Becker to H. Weyl." In Peckhaas 2005, 152–227.

Martin, Christopher A. 2003. "On Continuous Symmetries and the Foundations of Modern Physics." In Brading and Castellani 2003, 29–60.

Mielke, Eckehard W., and Friedrich W. Hehl. 1988. "Die Entwicklung der Eichtheorien: Marginalien zu deren Wissenschaftsgeschichte." In Deppert 1988, 191–230.

Moore, Walter. 1989. *Schrödinger: Life and Thought.* Cambridge: Cambridge University Press.

Narkiewicz, Władisław. 1988. "Hermann Weyl and the Theory of Numbers." In Deppert 1988, 51–60.

Nernst, W. 1922. "Zum Gültigkeitsbereich der Naturgesetze." *Die Naturwissenschaften* 10.

Newman, M.H.A. 1957. "Hermann Weyl, 1885–1955." *Biographical Memoirs of Fellows of the Royal Society London* 3:305–328.

Newton, Isaac. 1999. *The* Principia: *Mathematical Principles of Natural Philosophy*, tr. I. Bernard Cohen and Anne Whitman. Berkeley: University of California Press.

O'Raifeartaigh, Lochlainn, ed. 1972. *General Relativity: Papers in Honour of J. L. Synge.* Oxford: Clarendon Press.

———. ed. 1997. *The Dawning of Gauge Theory.* Princeton: Princeton University Press.

O'Raifeartaigh, Lochlainn, and Norbert Straumann. 2000. "Gauge Theory: Historical Origins and Some Modern Developments." *Reviews of Modern Physics* 72: 1–23.

Pais, Abraham. 1982. *"Subtle is the Lord. . .": The Science and the Life of Albert Einstein.* Oxford: Oxford University Press.

———. 1986. *Inward Bound.* Oxford: Oxford University Press.

Pauli, Wolfgang. 1979. *Wissenschaftlicher Briefwechsel.* Berlin: Springer-Verlag.

———. 1981. *Theory of Relativity*, tr. G. Field. New York: Dover.

Peckhaus, Volker. 1990. *Hilbertprogramm und Kritische Philosophie: Das Göttinger Modell interdisziplinärer Zusammenarbeit zwischen Mathematik und Philosophie.* Göttingen: Vandenhoeck und Ruprecht.

———. ed. 2005. *Oscar Becker und die Philosophie der Mathematik.* Munich: Wilhelm Fink Verlag.

Penrose, Roger. 1986. "Hermann Weyl, Space-Time, and Conformal Geometry." In Chandrasekharan 1986, 23–52.

Pesic, Peter. 1988–1989. "The Fields of Light." *St. John's Review* 38(3):1–16.

———. 1993. "Euclidean Hyperspace and Its Physical Significance." *Il Nuovo Cimento* 108B:1145–1153.

———. 1996. "Einstein and Spinoza: Determinism and Identicality Reconsidered." *Studia Spinozana* 12:195–203.

———. 2002. *Seeing Double: Shared Identities in Physics, Philosophy, and Literature.* Cambridge, MA: MIT Press.

———. 2003. *Abel's Proof: An Essay on the Sources and Meaning of Mathematical Unsolvability.* Cambridge, MA: MIT Press.

———. 2005. *Sky in a Bottle.* Cambridge, MA: MIT Press.

———. 2006. "Newton and the Mystery of the Major Sixth: A Transcription of His Manuscript 'Of Musick' with Commentary." *Interdisciplinary Science Reviews* 31:291–306.

———, ed. 2007. *Beyond Geometry: Classic Papers from Riemann to Einstein.* Mineola, NY: Dover.

Pesic, Peter, and Stephen P. Boughn. 2003. "The Weyl-Cartan Theorem and the Naturalness of General Relativity." *European Journal of Physics* 24:261–266.

Peterson, Mark A. 1979. "Dante and the 3-Sphere." *American Journal of Physics* 47:1031–1035.

Problèmes de Philosophie des Sciences. Archives de L'Institut International des Sciences Théoriques. Premièr Symposium—Bruxelles 1947. In *Bulletin de L'Académie Internationale des Sciences de la Nature,* Serie A. 7 fascicles. Paris: Hermann, 1948–1949.

Problems of Mathematics: Princeton University Bicentennial Conferences. 1946. Princeton: Princeton University.

Redhead, Michael. 2003. "The Interpretation of Gauge Symmetry." In Brading and Castellani 2003, 124–139.

Reich, Karin. 1992. "Levi-Civitasche Parallelverschiebung, affiner zusammenhang, Übertragungsprinzip: 1916/17–1922/23." *Archive for History of Exact Science* 44: 77–105.

Reichenbach, Hans. 1958. *The Philosophy of Space and Time,* tr. Maria Reichenbach and John Freund. New York: Dover.

Reid, Constance. 1996. *Hilbert.* New York: Copernicus.

Renn, J., J. Ritter, and T. Sauer, eds. 1999. *The Expanding Worlds of General Relativity.* Basel: Birkhäuser.

Ria, D. 2005. *L'unità fisico-matematica nel pensiero epistemologico di Hermann Weyl.* Lecce: Congedo Editore.

Riemann, Bernhard. 1919. *Über die Hypothesen, welche der Geometrie zu Grunde liegen,* ed. Hermann Weyl. Berlin: Springer. (Reprint, 1923)

Röller, Nils. 2002. *Medientheorie im epistemischen Übergang Hermann Weyls Philosophic der Mathematik und Naturwissenschaft und Ernst Cassirers Philosophie der symbolischen Formen in Wechselverhältnis.* Weimar: Verlag und Databank für Geisteswissenschaften.

Rowe, David. 1994. "The Philosophical Views of Klein and Hilbert." In Chikara et al. 1994, 187–202.

———. 1999. "The Göttingen Response to General Relativity and Emmy Noether's Theorem." In Gray 1999b, 189–223.

Ryckman, T. A. 1996. "Einstein *Agonists*: Weyl and Reichenbach on Geometry and the General Theory of Relativity." In Giere and Richardson 1996, 165–209.

———. 2003. "The Philosophical Roots of the Gauge Principle: Weyl and Transcendental Phenomenological Idealism." In Brading and Castellani 2003, 61–88.

Ryckman, Thomas. 2005. *The Reign of Relativity: Philosophy in Physics 1915–1925*. New York: Oxford University Press.

Saaty, Thomas L., and F. Joachim Weyl, eds. 1969. *The Spirit and the Uses of the Mathematical Sciences*. New York: McGraw-Hill.

Sauer, Tilmann. 1999. "The Relativity of Discovery." *Archive for History of Exact Sciences* 53:529–575.

Scheibe, Erhard. 1988. "Hermann Weyl and the Nature of Spacetime." In Deppert 1988, 61–82.

Schlick, Moritz. 1963. *Space and Time in Contemporary Physics: An Introduction to the Theory of Relativity and Gravitation*, tr. H. L. Brose. New York: Dover.

Scholz, Erhard. 1980. *Geschichte des Mannigfaltigkeitsbegriff von Riemann bis Poincaré*. Basel: Birkhäuser.

———. 1994. "Hermann Weyl's Contributions to Geometry, 1917–1923." In Chikara et al. 1994, 203–230.

———. 1995. "Hermann Weyl's 'Purely Infinitesimal Geometry.'" In Chatterji 1995, 1592–1603.

———. 1999a. "Weyl and the Theory of Connections." In Gray 1999b, 260–284.

———. 1999b. "The Concept of Manifold, 1850–1950." In James 1999, 25–64.

———. 2000. "Hermann Weyl on the Concept of Continuum." In Hendricks et al. 2000, 195–217.

———. 2001a. "Weyls Infinitesimalgeometrie, 1917–1925." In Scholz 2001b, 48–104.

———. ed. 2001b. *Hermann Weyl's Raum-Zeit-Materie and a General Introduction to His Scientific Work*. Basel: Birkhäuser.

———. 2004a. "Hermann Weyl's Analysis of the 'Problem of Space' and the Origin of Gauge Structures." *Science in Context* 17:165–197.

———. 2004b. "The Introduction of Groups into Quantum Theory." *Historia Mathematica* 33:440–490.

———. 2005a. "Philosophy as a Cultural Resource and Medium of Reflection for Hermann Weyl." *Revue de synthèse* 126:331–351.

———. 2005b. "Curved Spaces: Mathematics and Empirical Evidence, ca. 1830–1923." *Oberwolfach Reports* 4:3195–3198.

———. 2006. "Practice-Related Symbolic Realism in H. Weyl's Mature View of Mathematical Knowledge." In Ferreiros and Gray 2006, 291–310.

———. 2007. "The Changing Concept of Matter in H. Weyl's Thought, 1918–1930." In Hendricks et al., 281–305.

———. 2008. "Weyl Entering the 'New' Quantum Mechanical Discourse." In Joas et al., 2008, 2:253–271.

Schottky, W. 1921. "Das Kausalproblem der Quantentheorie als eine Grundfrage der modernen Naturforschung überhaupt." *Die Naturwissenschaften* 9.

Schrödinger, Erwin. 1922. "Über eine bemerkenswerte Eigenschaft der Quantenbahnen eines einzelnen Elektrons." *Zeitschrift für Physik* 12:13–23. (Translation in O'Raifeartaigh 1997, 87–90)

Schweber, S. S. 1994. *QED and the Men Who Made It: Dyson, Feynman, Schwinger, and Tomonaga*. Princeton: Princeton University Press.

———. 2000. *In the Shadow of the Bomb: Bethe, Oppenheimer, and the Moral Responsibility of the Scientist*. Princeton: Princeton University Press.

Schwinger, Julian. 1968. "Sources and Gravitons." *Physical Review* 173:1264–1272.

———. 1988. "Hermann Weyl and Quantum Kinematics." In Deppert 1988, 107–129.

Sciama, Dennis. 1969. *The Physical Foundations of General Relativity*. New York: Doubleday.

Shevelle, Steven K., ed. 2003. *The Science of Color*. Boston: Elsevier.

Sieroka, Norman. 2007. "Weyl's 'Agens Theory' of Matter and the Zurich Fichte." *Studies in the History and Philosophy of Science* 38:84–107.

———. 2008. "Hermann Weyl (1885–1955)." In Weber and Desmond 2008, 2:197–206.

Sigurdsson, Skúli. 1991. "Hermann Weyl, Mathematics and Physics, 1900–1927." Ph.D. diss., Harvard University.

———. 1994. "Unification, Geometry and Ambivalence: Hilbert, Weyl and the Göttingen Community." In Gavroglu et al. 1994, 355–367.

———. 1996. "Physics, Life, and Contingency: Born, Schrödinger, and Weyl in Exile." In Ash and Söllner 1996, 48–70.

———. 2001. "Journeys in Spacetime." In Scholz 2001b, 15–47.

———. 2007. "On the Road." In Gavroglu and Renn 2007, 149–157.

Slodowy, Peter. 1999. "The Early Development of the Representation Theory of Semisimple Lie Groups: A. Hurwitz, I. Schur, H. Weyl." *Jahresbericht der Deutschen Mathematiker-Vereinigung* 101:97–115.

Speiser, David. 1988. "*Gruppentheorie und Quantenmechanik*: The Book and Its Position in Weyl's Work." In Deppert 1988, 161–189.

Straumann, Norman. "Ursprünge der Eichtheorien." In Scholz 2001c, 138–160.

Struik, Dirk J. 1989. "Schouten, Levi-Civita, and the Emergence of Tensor Calculus." In Rowe and McCleary 1989, 99–105.

Suppes, Patrick, ed. 1973. *Space, Time, and Geometry*. Dordrecht: D. Reidel.

Tavel, M. A. 1971. "Noether's Theorem." *Transport Theory and Statistical Physics* 1:183–207.

Tonietti, Tino. 1988. "Four Letters of E. Husserl to H. Weyl and Their Context." In Deppert 1988, 343–384.

Van Atten, Mark, Dirk van Dalen, and Richard Tieszen. 2002. "The Phenomenology and Mathematics of the Intuitive Continuum." *Philosophia Mathematica* 10:203–226.

Van Dalen, Dirk. 1984. "Four Letters from Edmund Husserl to Hermann Weyl." *Husserl Studies* 1:1–12.

———. 1995. "Hermann Weyl's Intuitionistic Mathematics." *Bulletin of Symbolic Logic* 1:145–169.

———. 1999a, 2005. *Mystic, Geometer, and Intuitionist. The Life of L.E.J. Brouwer*. 2 vols. Oxford: Oxford University Press.

Vizgin, Vladimir. 1994. *Unified Field Theories in the First Third of the 20th Century*, tr. J. B. Barbour. Basel: Birkhäuser.

Weber, Michel, and William Desmond, Jr., eds. 2008. *Handbook of Whiteheadian Process Thought*. Frankfurt: Ontos-Verlag.

Weiner, C., ed. 1977. *History of Twentieth-Century Physics: Proceedings of the International School of Physics "Enrico Fermi,"* course 57, Varenna, Italy. New York: Academic Press.

Wells, Raymond O., ed. 1988. *The Mathematical Heritage of Hermann Weyl: Proceedings of the Symposium on the Mathematical Heritage of Hermann Weyl, held at Duke University, Durham, North Carolina, May 12–16, 1987*. Providence, RI: American Mathematical Society.

Weyl, Hermann. 1913. *Die Idee der Riemannschen Fläche*. Leipzig: Teubner. (Translation, Weyl 1964)

———. 1918a. *Das Kontinuum. Kritische Untersuchungen über die Grundlagen der Analysis*. Leipzig: Veit. (Translation, Weyl 1994)

———. 1918b. *Raum-Zeit-Materie*. Berlin: Springer. (Translation, Weyl 1952a)

———. 1918c. "Gravitation und Elektrizität." *Sitzungsberichte der Königlich Preussischen Akademie der Wissenschaften zu Berlin* 465–480; *WGA* 2:29–42. (Translations in Lorentz et al. 1923, 200–216; O'Raifeartaigh 1997, 24–37)

———. 1918d. "Reine Infinitesimalgeometrie." *Mathematische Zeitschrift* 2:384–411; *WGA* 2:1–28.

———. 1920a. "Elektrizität und Gravitation." *Physikalische Zeitschrift* 21:649–650; *WGA* 2:141–142.

———. 1920b. "Das Verhältnis der kausalen zur statistischen Betrachtungsweise in der Physik." *Schwiezer Medizinische Wochenschrift*; *WGA* 2:113–122.

———. 1921a. "Electricity and Gravitation." *Nature* 106:800–802; *WGA* 2:260–262.

———. 1921b. "Uber die neue Grundlagenkrise der Mathematik." *Mathematische Zeitschrift* 10:39–79, *WGA* 2:143–180. (Translation in Mancosu 1998, 86–121)

———. 1923. *Mathematische Analyse des Raumproblems. Vorlesungen gehalten in Barcelona und Madrid*. Berlin: Springer. (Reprint, 1963)

———. 1924a. "Massenträgheit und Kosmos. Ein Dialog." *Die Naturwissenschaften* 12:197–204; *WGA* 2:478–485.

———. 1924b. "Was ist Materie?" *Die Naturwissenschaften* 12:561–568, 585–593, 604–611; *WGA* 2:486–510.

———. 1926. "Modern Conceptions of the Universe." In *Encyclopædia Brittanica*, 11th ed., vol. 3 (supplement):908–911; not included in *WGA*.

———. 1927a. "Zeitverhaeltnisse im Kosmos, Eigenzeit, gelebte Zeit und metaphysische Zeit." In Brightman 1927, 54–58; not included in *WGA*.

———. 1927b. *Philosophie der Mathematik und Naturwissenschaft, Handbuch der Philosophie*, Abt. 2A. Munich: Oldenbourg. (Translation in Weyl 1949a, 2nd ed. 1966)

———. 1928. *Gruppentheorie und Quantenmechanik*. Leipzig: Hirzel. (Translation in Weyl 1950)

———. 1929. "Elektron und Gravitation." *Zeitschrift für Physik* 56:330–352; *WGA* 3:245–267. (Translation in O'Raifeartaigh 1997, 121–144)

———. 1931. *Die Stufen des Unendlichen*. Jena: Gustav Fischer.

———. 1932. *The Open World: Three Lectures on the Metaphysical Implications of Science*. New Haven: Yale University Press. (Reprinted 1989, Woodbridge, CT: Ox Bow Press)

———. 1934a. *Mind and Nature*. Philadelphia: University of Pennsylvania Press.

———. 1934b. "Universum und Atom." *Die Naturwissenschaften* 22:145–149; *WGA* 3:420–424.

———. 1935. "Emmy Noether 1882–1935." *Scripta Mathematica* 3:201–220; *WGA* 3:425–444.

———. 1938. "Symmetry." *Journal of the Washington Academy of Sciences*. 28: 253–271.

———. 1939. *The Classical Groups, Their Invariants and Representations*. Princeton: Princeton University Press. (2nd ed., 1946)

———. 1940. "The Mathematical Way of Thinking." *Science* 92:437–446; *WGA* 3:710–718.

———. 1943. "On Hodge's Theory of Harmonic Integrals." *Annals of Mathematics* 44:1–6; *WGA* 4:115–120.

———. 1944. "David Hilbert and His Mathematical Work." *Bulletin of the American Mathematical Society* 50:612–654; *WGA* 4:130–172.

———. 1946. "Bicentennial conference." ETH Archive, unpublished typed manuscripts Hs 91a:17 (version 1), Hs 91a:18 (version 2). (For further details and editorial information, see the note at the beginning of the "Princeton Bicentennial Lecture.")

———. 1947. "Discussion Remarks by Hermann Weyl. First Symposium of the Institut International des Sciences Théoriques. Brussels, September 1947." Unpublished typed manuscript in the ETH Archiv, Hs 91a:19. Weyl's remarks were included as part of the transcripts of discussions throughout the published proceedings of this symposium, *Problèmes de Philosophie* 1948–1949.

———. 1949a. *Philosophy of Mathematics and Natural Science*, tr. Olaf Helmer. Princeton: Princeton University Press. (2nd ed., 1966; reprint, 2009)

———. 1949b. "Wissenschaft als symbolische Konstruction des Menschen." *Eranos-Jahrbuch 1948*, 375–431; *WGA* 4:289–345.

———. 1950. *The Theory of Groups and Quantum Mechanics*, tr. H. P. Robertson. New York: Dover.

———. 1952a. *Space-Time-Matter*. 4th German ed., tr. Henry L. Brose. New York: Dover.

———. 1952b. *Symmetry*. Princeton: Princeton University Press. (Reprint, 1980)

———. 1953a. "Über den Symbolismus der Mathematik und mathematischen Physik." *Studium Generale* 6:219–228; *WGA* 4:527–536.

———. 1953b. "Universities and Science in Germany." *Mathematics Student* (Madras, India) 21:1–26; *WGA* 4:537–562.

Weyl, Hermann. 1954. "Address on the Unity of Knowledge delivered at the Bicentennial Conference of Columbia University"; *WGA* 4:623–630. (Also included in Ayoub 2004, 67–77)

———. 1955a. "Erkenntnis und Besinnung (Ein Lebensrückblick)." *Studia Philosophica* 15; *WGA* 4:631–649. (Translation in Saaty and Weyl 1969, 281–301)

———. 1955b. "Rückblick auf Zürich aus dem Jahre 1930." *Schweizerische Hochschulzeitung* 28:180–189; *WGA* 4:650–654.

———. 1955c. *Selecta*. Basel: Birkhäuser.

———. 1964. *The Concept of a Riemann Surface*, tr. Gerald R. Maclane. 3rd ed. Reading, MA: Addison-Wesley.

———. 1968. *Gesammelte Abhandlungen*, ed. K. Chandrasekharan. Berlin: Springer. Cited as *WGA*.

———. 1977. *Mathematische Analyse des Raumproblems/Was ist Materie?* Darmstadt: Wissenschaftliche Buchgesellschaft.

———. 1988. *Riemanns geometrische Ideen, ihre Auswirkung und ihre Verknüpfung mit der Gruppentheorie (1925)*, ed. K. Chandrasekharan. Berlin: Springer-Verlag.

———. 1994. *The Continuum*, tr. S. Pollard and T. Bole. New York: Dover.

Wheeler, John. 1962. *Geometrodynamics*. New York: Academic Press.

Wheeler, John Archibald. 1968. *Einsteins Vision: wie stehtes heute mit Einsteins Vision alles als Geometrie aufzufassen?* Berlin: Springer-Verlag.

Wheeler, John A. 1988. "Hermann Weyl and the Unity of Knowledge." In Deppert 1988, 469–503. (An adapted version of this essay had appeared in 1986 in *American Scientist* 74:366–375.)

Wheeler, John A., and Richard P. Feynman. 1949. "Classical Electrodynamics in Terms of Direct Interparticle Action." *Reviews of Modern Physics* 21:235–433.

Woit, Peter. 2006. *Not Even Wrong: The Failure of String Theory and the Search for Unity in Physical Law*. New York: Basic Books.

Woodward, William R. 1978. "From Association to Gestalt: The Fate of Hermann Lotze's Theory of Spatial Perception, 1846–1920." *Isis* 69:572–582.

Wu, T. T., and C. N. Yang. 1975. "Concept of Non-Integrable Phase Factors and Global Formulation of Gauge Fields." *Physical Review D* 12:3845–3857.

Yang, Chen Ning. 1986. "Hermann Weyl's Contribution to Physics." In Chandrasekharan 1986, 7–21.

Zisman, M. 1999. "Fibre Bundles, Fibre Maps." In James 1999, 605–629.

Acknowledgments ▣

I thank Michael Weyl for giving permission for the use of the writings of his father, Hermann Weyl, included in this volume. I also thank those who have given permission to reprint the following works included here: Hermann Weyl, *The Open World: Three Lectures on the Metaphysical Implications of Science*, © 1932 Yale University Press, reprinted by permission of Yale University Press; Hermann Weyl, *Mind and Nature*, © 1934 University of Pennsylvania Press, reprinted by permission of the University of Pennsylvania Press; Hermann Weyl, "Address at the Princeton Bicentennial Conference" and "Man and the Foundations of Science," ETH-Bibliothek, Archive und Nachlässe, unpublished manuscripts Hs. 91a:17, 18, 28, reproduced with the permission of the ETH-Bibliothek, Archive und Nachlässe; Hermann Weyl, "The Unity of Knowledge," in Hermann Weyl, *Gesammelte Abhandlungen*, ed. K. Chandrasekharan, vol. 4, pp. 623-630 (© 1968 Springer-Verlag), reprinted with the kind permission of Springer Science+Business Media; Hermann Weyl, "Insight and Reflection," from *The Spirit and the Uses of the Mathematical Sciences*, edited by Thomas L. Saaty and F. Joachim Weyl (© 1969 McGraw-Hill Book Company, New York), pp. 281-301, reproduced with the permission of The McGraw-Hill Companies and of the Schweizerische Philosophische Gesellschaft, in whose journal *Studia Philosophica*, vol. XV (1955), this paper originally appeared as Hermann Weyl, "Erkenntnis und Besinnung (Ein Lebensrückblick)." I thank Nils Röller, Erhard Scholz, Norman Sieroka, Skúlli Sigursson, and Michael Gasser (ETH-Bibliothek) for their kind advice and help in locating manuscripts of Hermann Weyl.

For gracious permission to print the photographs included here, I thank Michael Weyl, and I thank Thomas Weyl, Peter Weyl, Annemarie Weyl Carr, and Thomas Ryckman for their invaluable help in locating images. Unless listed below, all photographs come from the Weyl family collections. For permission to publish the specific photos listed here, I thank: (1) Mathematisches Forschungsinstitut Oberwolfach gGmbH (MFO); (2) Archives of the Mathematisches Forschungsinstitut Oberwolfach; (3), (5) courtesy of the Niedersächsische Staats- und Universitätsbibliothek Göttingen, Abteilung für Handschriften und Seltene Drucke, (3) Sammlung Voit: H. Weyl and (5) Cod. Ms. D. Hilbert 754: Nr. 98; (12), (14), (18), (22) from *Hermann Weyl: 1885–1985; Centenary Lectures delivered by C. N. Yang, R. Penrose, and A. Borel at the Eidgenössische Technische Hochschule Zürich*, edited by K. Chandrasekharan (© 1968 Springer-Verlag), reproduced with the kind permission of Springer Science+Business Media; (17) courtesy AIP Emilio Segré Visual Archives.

Index ▣

Index

Racah, Giulio (1909–1965), 228n31
rainbow, 84
Ramanujan, Srinivasa (1887–1920), 172
Ranke, Leopold von (1795–1886), 202
ratio, anharmonic, 90
realism, 96–97, 107–108, 216
Reidemeister, Kurt (1893–1971), 200–201, 203
relativity, 134; Galilean, 22, 122; general theory of, 2, 10, 20–24, 45, 132–134, 168, 210–211, 224n7, 228n30; of magnitude, 3, 22; special theory of, 41–44; of time, 32–33
religion, 16, 35, 37, 168, 192, 195, 209, 211
Riemann, Georg Bernhardt (1826–1866), 133, 172, 206–207, 211, 224n13; Riemann surfaces, 166, 237n10, 240n4
Riesz, Frigyes (1880–1956), 164
Ritz-Rydberg combination principle, 142
Roemer, Ole (1644–1710), 114, 130
Runge, Carl (1856–1927), 167
Russell, Bertrand (1872–1970), 76, 169, 185–186
Ryckman, Thomas, 225n15, 226n23, 228n32

Scheler, Max (1874–1928), 189
Scherrer, Paul (1890–1969), 3
Scherrer, Willy, 3, 168
Schlick, Moritz (1882–1936), 108, 234n1
Schmidt, Erhard (1876–1959), 163–164
Scholz, Erhard , 224n13, 225n17, 226n19
Scholz, Wilhelm von (1874–1969), 84, 234n2
Schrödinger, Anny (1896–1965), 3, 5, 224n10
Schrödinger, Erwin (1887–1961), 3, 5–6, 142, 169, 225nn17 and 18
Schur, Issai (1875–1941), 167–168
Schwarz, Hermann Amandus (1843–1921), 163–165
Schweitzer, Albert (1875–1965), 197
Schwinger, Julian (1918–1994), 11, 228n30
scientific concepts and theories, 109–121, 114–116; future of, 196
sense perception, 13, 83–103
sets, theory of, 33, 76, 186
Shaftesbury, Anthony Ashley Cooper, Third Earl of (1671–1713), 37

Shakespeare, William (1564–1616), 5
Sieroka, Norman, 224n13, 226n19, 227n27
Sigurdsson, Skúli, 223n6, 226nn19 and 20, 229n34
simultaneity, 42–43, 129–133, 146
sky, color of, 61, 144
Slater, John (1900–1976), 11, 228n31
Snell, Willebrord (1580–1626), 55
snowflakes, 17–18
Sommerfeld, Arnold (1868–1951), 223n4
sound, perception of, 86, 92–93
space, 41, 116; absolute, 29, 44–45; as form of intuition, 99, 101; structure of, 133–134
space-time, 3–5, 8, 41–44, 231n2
Spinoza, Benedict de (1632–1951), 12, 26, 38, 231n1
Spitteler, Carl Friedrich Georg (1845–1924), 47, 232n8
"standard model" of particle physics, 87
statistical aspects of physics, 5–6, 61–62, 143–150; and thermodynamics, 164–165
Stern-Gerlach experiment, 64, 146, 233n15
Stoics, 44–45, 232n5
subjectivity, 31, 104, 134
Switzerland, 2, 13, 162
symbols, 12–13, 15, 104, 119, 137, 184, 195–203
symmetry, 7, 11, 228n32, 229nn40 and 41
Szegő, Gabor (1895–1985), 171

Takagi, Teiji (1875–1960), 167
TCP theorem, 11, 17–18, 228n32
tertium non datur [law of the excluded middle], 75
theology, 39–40, 44, 50, 125, 192
3-sphere, 231n1
time, 8–10, 15, 29–33; absolute, 29; as form of intuition, 99; and Heidegger, 192; lived, 32, 192; metaphysical, 32–33; "now" in, 33; proper, 31, 207; reversibility of, 62. *See also* causality; future; now; past, nature of; simultaneity
topology, 6, 166

ultra-violet, 93; "catastrophe," 235n17
unified field theories, 20–24, 230n3

260